AF541608

TEXT BOOK
OF
INTEGRAL CALCULUS

DPH MATHEMATICS SERIES

TEXT BOOK OF INTEGRAL CALCULUS

By

A.K. Sharma

DISCOVERY PUBLISHING HOUSE
NEW DELHI-110002

First Published - 2005

Reprinted - 2019

ISBN: 978-81-7141-968-5

Textbook of Integral Calculus

Published by:

DISCOVERY PUBLISHING HOUSE PVT. LTD.
4383/4B, Ansari Road, Darya Ganj
New Delhi-110 002 (India)
Phone: +91-11-23279245, 43596064-65
Fax: +91-11-23253475
E-mail: discoverypublishinghouse@gmail.com
sales@discoverypublishinggroup.com
web: www.discoverypublishinggroup.com

Printed at:
Infinity Imaging Systems
Delhi

Preface

This book "Text Book of Integral Calculus" has been specially written to meet the requirements of B.A./B.Sc., students of all Indian Universities.

The subject matter has been discussed in such a simple way that the students will find no difficulty to understand it. The proof of various theorems and examples has been given with minute details. Each chapter of this book contains complete theory and large number of solved examples. Sufficient problems have also been selected from various Indian Universities.

I hope this book warmly received by the students and teachers.

The author wish to express their thanks to the publisher M/s Discovery Publishing House, New Delhi to bringing out this book in the present nice form.

A.K. Sharma

CONTENTS

1

INTEGRATION OF TRIGONOMETRIC FUNCTIONS

1.1. INTEGRATION OF $\sin^m x$ AND $\cos^m x$

Case I: *When m is an odd positive integer.*

Let m = 2n + 1. Then

(i) $\int \sin^m x\,dx = \int \sin^{2n+1} x\,dx = \int \sin^{2n} x.\sin x\,dx$

$= \int (\sin^2 x)^n \sin x\,dx = \int (1-\cos^2 x)^n.\sin x\,dx$

$= -\int (1-t^2)^n\,dt$, putting cos x = t, so that – sin x dx = dt.

Now $(1 - t^2)^n$ can be expanded in powers of t by the binomial theorem and then term by term integration will be performed.

Thus $\int \sin^{2n+1} x\,dx$

$$= -\int \left[1 - nt^2 + \frac{n(n-1)}{2!}t^4 - ... + (-1)^n t^{2n}\right] dt$$

$$= -\left[t - n\frac{t^3}{3} + \frac{n(n-1)}{2!}\cdot\frac{t^5}{5} - ... + \frac{(-1)^n.t^{2n+1}}{2n+1}\right]$$

$$= -\left[\cos x - \frac{n}{3}\cos^3 x + \frac{n(n-1)}{5.2!}\cos^5 x - ... + \frac{(-1)^n}{(2n+1)}\cos^{2n+1} x\right].$$

(ii) Similarly, $\int \cos^m x\,dx = \int \cos^{2n+1} x\,dx$, [∵ m = 2n + 1]

$= \int \cos^{2n} x \cos x\,dx = \int (1-\sin^2 x)^n \cos x\,dx$, **(Note)**

$= \int (1-t^2)^n\,dt$, putting sin x = t and cos x dx = dt.

This integral can now be easily evaluated by expanding $(1-t^2)^n$ by the binomial theorem.

Case II: *When m is an even positive integer.*

(a) *When m is small,* the integration can be done by transforming the given integrand into a sum of cosines of multiles of x by using the following trigonometrical formulae

$$\cos^2 x = \frac{1}{2}(1+\cos 2x),\ \sin^2 x = \frac{1}{2}(1-\cos 2x).$$

(b) *When m is large,* we apply De-Moivre's theorem as explained below:

Let $\cos x + i \sin x = y.$

Then $\cos x - i \sin x = y^{-1} = \frac{1}{y}.$

Therefore $2\cos x = y + y^{-1} = y + \left(\frac{1}{y}\right)$

and $2i \sin x = y - y^{-1} = y - \left(\frac{1}{y}\right).$

Also $y^n + \left(\frac{1}{y^n}\right) = (\cos x + i \sin x)^n + (\cos x - i \sin x)^n$

$= (\cos nx + i \sin nx) + (\cos nx - i \sin nx) = 2 \cos nx,$

and $y^n = \left(\frac{1}{y^n}\right) = 2i \sin nx.$

Now $2^m \cos^m x = \left\{y + \left(\frac{1}{y}\right)\right\}^m$. Applying binomial theorem, we have

$$2^m \cos^m x = y^m + {}^m c_1 y^{m-1} \cdot \frac{1}{y} + {}^m c_2 y^{m-2} \cdot \frac{1}{y^2} + \ldots + m \cdot \frac{y}{y^{m-1}} + \frac{1}{y^m}$$

$$= y^m + m y^{m-2} + \frac{m(m-1)}{1.2} y^{m-4} + \ldots + \frac{m}{y^{m-2}} + \frac{1}{y^m}$$

$$= \left(y^m + \frac{1}{y^m}\right) + m\left(y^{m-2} + \frac{1}{y^{m-2}}\right) + \ldots$$

$= 2 \cos mx + m.2\cos (m - 2) x + \ldots$

Thus $\int 2^m \cos^m x\,dx = 2\int \cos mx\,dx + 2\int m\cos(m-2)x\,dx + \ldots$

$$= \frac{2}{m}\sin mx + \frac{2m}{m-2}\sin(m-2)x + \ldots$$

or $\int \cos^m x\,dx = \frac{1}{2^{m-1}}\left[\frac{1}{m}\sin mx + \frac{m}{m-2}\sin(m-2)x + \ldots\right].$

Similarly $(2i \sin x)^m = \left[y - \left(\frac{1}{y}\right)\right]^m$.

Expanding by binomial theorem and simplifying, we have

$2^m i^m \sin^m x = y^m - {}^mc_1\, y^{m-2} + {}^mc_2 y^{m-4} - \ldots$

$$+{}^mc_2 \frac{1}{y^{m-4}} - \frac{m}{y^{m-2}} + \frac{1}{y^m},$$ [Note that m is even]

$$= \left(y^m + \frac{1}{y^m}\right) - m\left(y^{m-2} + \frac{1}{y^{m-2}}\right) + \frac{m(m+1)}{2!}\left(y^{m-4} + \frac{1}{y^{m-4}}\right) - \ldots$$

$$= 2\cos mx - m\{2\cos(m-2)x\} + \frac{m(m-1)}{2!}\{2\cos(m-4)x\} - \ldots$$

Hence $\int \sin^m x\,dx$

$$= \frac{1}{2^{m-1}(-1)^{m/2}}\int[\cos mx - m\cos(m-2)x + \ldots]dx.$$

$[\because i^m = (i^2)^{m/2} = (-1)^{m/2}]$

1.2. INTEGRATION OF $\sin^m x \cos^n x$

Case I: *If m is odd, put cos x = t; If n is odd, put sin x = t.*

If both m and n are odd, put either sin x or cos x equal to t and integrate.

Case II: *When (m + n) is a negative even integer.*

Convert the given integral in terms of tan x and sec x and put tan x = t. Then expand by binomial theorem, if necessary, and integrate term by term.

Case III: *When both m and n are even integers.*

(i) If m and n are small even integers, then convert $\sin^m x \cos^n x$ in terms of multiples of angles by using the formulae

$$\cos^2 x = \frac{1}{2}(1 + \cos 2x),$$

$$\sin^2 x = \frac{1}{2}(1 - \cos 2x),\ \sin x \cos x = \frac{1}{2}\sin 2x,$$

and $2\cos x \cos y = \cos(x + y) + \cos(x - y)$.

(ii) When m and n are large, we make use of De-Moivre's theorem as explained in 4.1 Case II (b), term by term integration yields the desired result.

1.3. INTEGRATION OF $\frac{1}{(a+b\cos x)}$

We have $I=\int \frac{dx}{a+b\cos x}$

$$=\int \frac{dx}{a\left(\cos^2\frac{1}{2}x+\sin^2\frac{1}{2}x\right)+b\left(\cos^2\frac{1}{2}x-\sin^2\frac{1}{2}x\right)} \quad \textbf{(Note)}$$

$$=\int \frac{dx}{(a+b)\cos^2\frac{1}{2}x+(a-b)\sin^2\frac{1}{2}x}$$

$$=\int \frac{\sec^2\frac{1}{2}x\,dx}{a+b+(a-b)\tan^2\frac{1}{2}x},$$

dividing the numerator and the denominator by $\cos^2\frac{1}{2}x$.

Now putting $\tan\frac{1}{2}x=t$

so that $\frac{1}{2}\sec^2\frac{1}{2}x\,dx=dt$, we get

$$I=2\int \frac{dt}{(a+b)+(a-b)t^2} \quad ...(1)$$

Now two cases arise viz, a > b or a < b.

Case I: a > b.

In this case, we have from (1)

$$I=\frac{2}{a-b}\int \frac{dt}{t^2+k^2}, \quad \left[\text{putting } \frac{a+b}{a-b}=k^2\right]$$

$$=\frac{2}{(a-b)}\cdot\frac{1}{k}\tan^{-1}\frac{t}{k}=\frac{2}{(a-b)\sqrt{\left(\frac{a+b}{a-b}\right)}}\tan^{-1}\left\{\frac{t}{\sqrt{\left(\frac{a+b}{a-b}\right)}}\right\}$$

$$=\frac{2}{\sqrt{(a^2-b^2)}}\tan^{-1}\left\{\sqrt{\left(\frac{a-b}{a+b}\right)}\cdot t\right\}$$

$$= \frac{2}{\sqrt{(a^2 - b^2)}} \tan^{-1}\left[\sqrt{\left\{\frac{a-b}{a+b}\right\}} \tan\frac{x}{2}\right].$$

Case II: a < b.

In this case, from (1) we write

$$I = 2\int \frac{dt}{(a+b)-(b-a)t^2} = \frac{2}{b-a}\int \frac{dt}{\left\{\frac{(b+a)}{(b-a)}\right\} - t^2}$$

$$= \frac{2}{a-b} \cdot \frac{1}{2\sqrt{\left\{\frac{(b+a)}{(b-a)}\right\}}} \cdot \log \frac{\sqrt{\left\{\frac{(b+a)}{b-a}\right\}} + t}{\sqrt{\left\{\frac{(b+a)}{(b-a)}\right\}} - t}$$

$$= \frac{1}{\sqrt{(b^2 - a^2)}} \cdot \log \frac{\sqrt{(b+a)} + t\sqrt{(b-a)}}{\sqrt{(b+a)} - t\sqrt{(b-a)}}$$

$$= \frac{1}{\sqrt{(b^2 - a^2)}} \cdot \log \frac{\sqrt{(b+a)} + \sqrt{(b-a)} \tan\frac{1}{2}x}{\sqrt{(b+a)} - \sqrt{(b-a)} \tan\frac{1}{2}x}.$$

1.4. INTEGRATION OF $\frac{1}{(a + b \sin x)}$

We have $\int \frac{dx}{(a + b \sin x)}$

$$= \int \frac{dx}{a\left(\cos^2\frac{1}{2}x + \sin^2\frac{1}{2}x\right) + b\left(2\sin\frac{1}{2}x \cos\frac{1}{2}x\right)}$$

$$= \int \frac{\sec^2\frac{1}{2}x\,dx}{a\left(1 + \tan^2\frac{1}{2}x\right) + b\left(2\tan\frac{1}{2}x\right)},$$ dividing Nr. & Dr. by $\cos^2\frac{1}{2}x$

$$= 2\int \frac{dt}{(at^2 + 2bt + a)},$$

putting $\tan\frac{1}{2}x = t$,

so that $\frac{1}{2}\sec^2\frac{1}{2}x\,dx = dt$

$$= \frac{2}{a}\int \frac{dt}{t^2 + \left(\frac{2b}{a}\right)t + 1} = \frac{2}{a}\int \frac{dt}{\left\{t + \left(\frac{b}{a}\right)\right\}^2 + 1 - \left(\frac{b^2}{a^2}\right)}$$

$$= \frac{2}{a}\int \frac{dt}{\left\{t + \left(\frac{b}{a}\right)\right\}^2 + \left\{\frac{(a^2 - b^2)}{a^2}\right\}}. \qquad ...(1)$$

Now two cases arise viz. $a > b$ *or* $a < b$.

Case I: a > b, *i.e.,* $\left\{\frac{(a^2 - b^2)}{a^2}\right\}$ **is positive.**

Then

$$\int \frac{dx}{a + b\sin x} = \frac{2}{a}\cdot\frac{a}{\sqrt{(a^2 - b^2)}}\cdot\tan^{-1}\left[\frac{\left\{t + \left(\frac{b}{a}\right)\right\}}{\sqrt{\left\{\frac{(a^2 - b^2)}{a^2}\right\}}}\right], \text{ from (1)}$$

$$= \frac{2}{\sqrt{(a^2 - b^2)}}\tan^{-1}\left\{\frac{at + b}{\sqrt{(a^2 - b^2)}}\right\}$$

$$= \frac{2}{\sqrt{(a^2 - b^2)}}\tan^{-1}\left\{\frac{a\tan\frac{1}{2}x + b}{\sqrt{(a^2 - b^2)}}\right\}.$$

Case II: a < b, *i.e.,* $\left\{\frac{(a^2 - b^2)}{a^2}\right\}$ **is negative.**

Then $\int \frac{dx}{a + b\sin x} = \frac{2}{a}\int \frac{dt}{\left\{t + \left(\frac{b}{a}\right)\right\}^2 - \left\{\frac{(b^2 - a^2)}{a^2}\right\}}$, from (1)

$$= \frac{2}{a}\cdot\frac{1}{2}\sqrt{\left\{\frac{a^2}{(b^2 - a^2)}\right\}}\log\left[\frac{\left\{t + \left(\frac{b}{a}\right)\right\} - \sqrt{\left\{\frac{(b^2 - a^2)}{a^2}\right\}}}{\left\{t + \left(\frac{b}{a}\right)\right\} + \sqrt{\left\{\frac{(b^2 - a^2)}{a^2}\right\}}}\right]$$

$$=\frac{1}{\sqrt{(b^2-a^2)}}\log\left\{\frac{at+b-\sqrt{(b^2-a^2)}}{at+b+\sqrt{(b^2-a^2)}}\right\}$$

$$=\frac{1}{\sqrt{(b^2-a^2)}}\log\left\{\frac{a\tan\left(\frac{1}{2}x\right)+b-\sqrt{(b^2-a^2)}}{a\tan\left(\frac{1}{2}x\right)+b+\sqrt{(b^2-a^2)}}\right\}.$$

1.5. TO EVALUATE $\int\frac{dx}{a\sin x+b\cos x}$

The given integral

$$=\int\frac{dx}{a\left(2\sin\frac{1}{2}x\cos\frac{1}{2}x\right)+b\left(\cos^2\frac{1}{2}x-\sin^2\frac{1}{2}x\right)}$$

$$=\int\frac{\sec^2\left(\frac{1}{2}x\right)dx}{2a\tan\frac{1}{2}x+b\left(1-\tan^2\frac{1}{2}x\right)},\text{ dividing Nr. and Dr. by }\cos^2\frac{1}{2}x$$

$$=\int\frac{2\,dt}{2at+b-bt^2},\text{ putting }\tan\frac{1}{2}x=t\text{ so that }\frac{1}{2}\sec^2\frac{1}{2}x\,dx=dt$$

$$=\frac{2}{b}\int\frac{dt}{1-t^2+\left(\frac{2a}{b}\right)t}=\frac{2}{b}\int\frac{dt}{1-\left\{t^2-\left(\frac{2a}{b}\right)t\right\}}$$

$$=\frac{2}{b}\int\frac{dt}{\left\{1+\left(\frac{a^2}{b^2}\right)\right\}-\left\{t-\left(\frac{a}{b}\right)\right\}^2}$$

$$=\frac{2}{b}\int\frac{dt}{\left\{\frac{(b^2+a^2)}{b^2}\right\}-\left\{t-\left(\frac{a}{b}\right)\right\}^2}$$

$$=\frac{2}{b}\cdot\frac{1}{2}\sqrt{\left\{\frac{b^2}{(b^2+a^2)}\right\}}\log\left[\frac{\sqrt{\left\{\frac{(b^2+a^2)}{b^2}\right\}}+\left\{t-\left(\frac{a}{b}\right)\right\}}{\sqrt{\left\{\frac{(b^2+a^2)}{b^2}\right\}}-\left\{t-\left(\frac{a}{b}\right)\right\}}\right]$$

$$= \frac{1}{\sqrt{(b^2+a^2)}} \log\left[\frac{\sqrt{(b^2+a^2)}-a+bt}{\sqrt{(b^2+a^2)}+a-bt}\right]$$

$$= \frac{1}{\sqrt{(b^2+a^2)}} \log\left[\frac{\sqrt{(b^2+a^2)}-a+b\tan\frac{1}{2}x}{\sqrt{(b^2+a^2)}+a-b\tan\frac{1}{2}x}\right].$$

1.6. TO EVALUATE $\int \frac{dx}{a+b\cos x+c\sin x}$

The given integral

$$I = \int \frac{dx}{a\left(\cos^2\frac{1}{2}x+\sin^2\frac{1}{2}x\right)+b\left(\cos^2\frac{1}{2}x-\sin^2\frac{1}{2}x\right)+2c\sin\frac{1}{2}x\cos\frac{1}{2}x}$$

$$= \int \frac{dx}{(a-b)\sin^2\frac{1}{2}x+2c\sin\frac{1}{2}x\cos\frac{1}{2}x+(a+b)\cos^2\frac{1}{2}x}.$$

Now dividing the numerator and the denominator by $\cos^2\frac{1}{2}x$, we get

$$I = \int \frac{\sec^2\frac{1}{2}x\,dx}{(a-b)\tan^2\frac{1}{2}+2c\tan\frac{1}{2}x+(a+b)}$$

$$= \frac{1}{(a-b)}\int \frac{\sec^2\frac{1}{2}x\,dx}{\tan^2\frac{1}{2}x+\left\{\frac{2c}{(a-b)}\right\}\tan\frac{1}{2}x+\left\{\frac{(a+b)}{(a-b)}\right\}}$$

$$= \frac{2}{a-b}\int \frac{dt}{t^2+\left\{\frac{2c}{(a-b)}\right\}t+\left\{\frac{(a+b)}{(a-b)}\right\}}$$

putting $\tan\frac{1}{2}x = t$, $\frac{1}{2}\sec^2\frac{1}{2}x\,dx = dt$.

The integral can now be evaluated by the methods already discussed.

1.7. INTEGRATION OF $\frac{P\cos x+Q\sin x+R}{a\cos x+b\sin x+c}$

To integrate such a fraction we express the numerator in the form Numerator = A (Deno.) + B (diff. coeff. of Deno.) + C, where A, B and C are constants.

Thus, we write P cos x + Q sin x + R

= A (a cos x + b sin x + c) + B (– a sin x + b cos x) + C.

Comparing the coefficients of cos x, sin x and constant terms on both sides, we have

$$P = Aa + Bb \quad ...(1)$$

$$Q = Ab - Ba \quad ...(2)$$

$$R = Ac + C. \quad ...(3)$$

Whence $A = \dfrac{Pa + Qb}{b^2 + a^2}$,

$$B = \frac{Pb - Qa}{b^2 + a^2}$$

and $C = R - \dfrac{(Pa + Qb)c}{b^2 + a^2}$.

Now writing the numerator in the form mentioned above, the given integral becomes

$$I = \int \frac{A(\text{Dr.}) + B(\text{diff. coeff. of Dr.}) + C}{\text{Dr.}} dx$$

$$= A\int dx + B\int \frac{-a \sin x + b \cos x}{a \cos x + b \sin x + c} + C\int \frac{dx}{a \cos x + b \sin x + c}$$

$$= Ax + B \log (a \cos x + b \sin x + c) + C \int \frac{dx}{a \cos x + b \sin x + c}.$$

The last integral can now be evaluated by the methods discussed earlier.

Cor.: $\displaystyle\int \frac{P \cos x + Q \sin x}{a \cos x + b \sin x} dx$ is a particular case of 4.7. Here neither the numerator nor the denominator contains a constant term. So here we express the numerator in the form

Nr. = A (Dr.) + B (diff. coeff. of Dr.).

1.8. INTEGRATION OF $\dfrac{1}{(a + b \tan x)}$

Let $I = \displaystyle\int \frac{dx}{a + b \tan x} = \int \frac{\cos x \, dx}{a \cos x + b \sin x}$.

Now we express the numerator cos x in the form

Nr. = A (Dr.) + B (diff. coeff. of Dr.)

i.e., cos x = A (a cos x + b sin x) + B (– a sin x + b cos x).

Equating the coefficients of cos x and sin x on both sides, we get

Aa + Bb = 1, and Ab – Ba = 0.

Solving these equations, we get

$$A=\frac{a}{\left(a^2+b^2\right)} \text{ and } B=\frac{b}{\left(a^2+b^2\right)}.$$

$$\text{Now } I=\int\frac{A(a\cos x+b\sin x)+B(-a\sin x+b\cos x)}{a\cos x+b\sin x}dx$$

$$=A\int dx+B\int\frac{-a\sin x+b\cos x}{a\cos x+b\sin x}dx$$

$$= Ax + B\log(a\cos x + b\sin x)$$

$$=\left\{\frac{a}{a^2+b^2}\right\}x+\frac{b}{a^2+b^2}\log(a\cos x+b\sin x).$$

MISCELLANEOUS EXAMPLES

Example 1:

Evaluate $\int \sin^6 x\,dx$.

Solution:

Let cos x + i sin x = y.

Then cos x – i sin x = y^{-1}

$\therefore$ 2i sin x = y – y^{-1}, or $2^6 i^6 \sin^6 x = (y - y^{-1})^6$

$$\text{or } -64\sin^6 x = y^6 - {}^6c_1 y^5\cdot\frac{1}{y} + {}^6c_2 y^4\cdot\frac{1}{y^2} - {}^6c_3 y^3\cdot\frac{1}{y^3}$$

$$+{}^6c_4 y^2\cdot\frac{1}{y^4} - {}^6c_5 y\cdot\frac{1}{y^5} + {}^6c_6\frac{1}{y^6}0, \qquad \text{[by binomial theorem]}$$

$$=\left(y^6+\frac{1}{y^6}\right)-6\left(y^4+\frac{1}{y^4}\right)+25\left(y^2+\frac{1}{y^2}\right)-20.$$

Hence – 4 $\sin^6$ x = 2 cos 6x – 6 . 2 cos 4x + 15 . 2cos 2x – 20,

$$\left[\because y^n+\left(\frac{1}{y^n}\right)=2\cos nx\right]$$

$$\text{or } \sin^6 x = -\frac{1}{32}\,[\cos 6x - 6\cos 4x + 15\cos 2x - 10].$$

Now integrating, we have $\int \sin^6 x\,dx$

$$=\frac{1}{32}\left[\int \cos 6x\,dx - 6\int \cos 4x\,dx + 15\int \cos 2x\,dx - \int 10\,dx\right]$$

$$=-\frac{1}{32}\left[\frac{\sin 6x}{6} - 6\frac{\sin 4x}{4} + \frac{15\sin 2x}{2} - 10x\right].$$

Example 2:

Evaluate $\int \cos^6 x\,dx$.

Solution:

Let $\cos x + i \sin x = y$.

Then $\cos x - i \sin x = y^{-1}$.

$\therefore\ 2 \cos x = (y + y^{-1})$, or $2^6 \cos^6 x = (y + y^{-1})^6$

$$\text{or } 2^6 \cos^6 x = \left[y+\frac{1}{y}\right]^6 = y^6 + {}^6c_1 y^5\cdot\frac{1}{y} + {}^6c_2 y^4\cdot\frac{1}{y^2} + {}^6c_3 y^3\cdot\frac{1}{y^3}$$

$$+{}^6c_4 y^2\cdot\frac{1}{y^4} + {}^6c_5 y\cdot\frac{1}{y^5} + {}^6c_6\frac{1}{y^6}$$

$$=\left(y^6+\frac{1}{y^6}\right) + {}^6c_1\left(y^4+\frac{1}{y^4}\right) + {}^6c_2\left(y^2+\frac{1}{y^2}\right) + 20$$

$= 2 \cos 6x + 6 . 2 \cos 4x + 15 . 2 \cos 2x + 20.$

Hence $32 \cos^6 x = \cos 6x + 6\cos 4x + 15 \cos 2x + 10.$

$$\therefore\ \int \cos^6 x\,dx = \frac{1}{32}\left[\frac{1}{6}\sin 6x + \frac{6\sin 4x}{4} + \frac{15\sin 2x}{2} + 10x\right].$$

Example 3:

Evaluate $\int \cos^8 x\,dx$.

Solution:

Let $\cos x + i \sin x = y$.

Then $\cos x - i \sin x = y^{-1}$

so that $2 \cos x = y + y^{-1}$, or $2^8 \cos^8 x = [y + y^{-1}]^8$

or $2^8 \cos^8 x = y^8 + {}^8c_1 y^6 + {}^8c_2 y^4 + {}^8c_3 y^2 + {}^8c_4$

$$+{}^8c_5\frac{1}{y^2} + {}^8c_6\frac{1}{y^4} + {}^8c_7\frac{1}{y^6} + \frac{1}{y^8}$$

$$=\left(y^8+\frac{1}{y^8}\right)+8\left(y^6+\frac{1}{y^6}\right)+28\left(y^4+\frac{1}{y^4}\right)+56\left(y^2+\frac{1}{y^2}\right)+70$$

$= 2 \cos 8x + 8.2 \cos 6x + 28.2 \cos 4x + 56.2\cos 2x + 70.$

$\therefore \int \cos^8 x\,dx$

$$=\frac{1}{128}\int(\cos 8x+8\cos 6x+28\cos 4x+56\cos 2x+35)\,dx$$

$$=\frac{1}{128}\left(\frac{\sin 8x}{8}+8\frac{\sin 6x}{6}+2.\frac{\sin 4x}{4}+56.\frac{\sin 2x}{2}+35x\right).$$

Example 4:

Integrate $\sin^2 x \cos^3 x$.

Solution:

Here the power of cos x being odd,

we put sin x = t,

so that cos x dx = dt.

Thus $\int \sin^2 x\cos^3 x\,dx = \int \sin^2 x\cos^2 x\cos x\,dx$

$= \int \sin^2 x\left(1-\sin^2 x\right)\cos x\,dx$

$= \int t^2\left(1-t^2\right)dt.$ [Putting sin x = t so that cos x dx = dt]

$= \int\left(t^2-t^4\right)dt = \frac{1}{3}t^3-\frac{1}{5}t^5 = \frac{1}{3}\sin^3 x-\frac{1}{5}\sin^5 x.$

Example 5:

Evaluate:

(i) $\int \sin^4 x\cos^3 x\,dx$

(ii) $\int \sin^3 x\cos^2 x\,dx$.

Solution:

(i) Let $I = \int \sin^4 x\cos^3 x\,dx = \int \sin^4 x\cos^2 x\cos x\,dx$

$= \int \sin^4 x\left(1-\sin^2 x\right)\cos x\,dx.$

Put sin x = t

so that cos x dx = dt.

Then

$$I=\int t^4(1-t^2)dt=\int(t^4-t^6)dt=\frac{t^5}{5}-\frac{t^7}{7}=\frac{\sin^5 x}{5}-\frac{\sin^7 x}{7}$$

(ii) Let $I=\int \sin^3 x\cos^2 x\,dx=\int \sin^2 x\cos^2 x\sin x\,dx$

$$=\int(1-\cos^2 x)\cos^2 x\ \sin x\,dx.$$

Put cos x = t

so that – sin x dx = dt.

Then $I=\int(1-t^2)t^2(-dt)=-\int(t^2-t^4)dt$

$$=-\frac{t^3}{3}+\frac{t^5}{5}=-\frac{\cos^3 x}{3}+\frac{\cos^5 x}{5}.$$

Example 6:

Integrate $\sin^5 x\cos^4 x$.

Solution:

Here the power of sin x being odd, we put cos x = t.

$$\therefore\ \int \sin^5 x\cos^4 x\,dx=\int \sin^4 x\cos^4 x\ \sin x\,dx$$

$$=\int(1-\cos^2 x)^2\cos^4 x\sin x\,dx$$

$$=-\int(1-t^2)^2 t^4dt,\qquad [\because\ \cos x=t \text{ and } -\sin x\,dx=dt]$$

$$=-\int(1-2t^2+t^4)t^4\,dt=\int(-t^4+2t^6-t^8)dt$$

$$=-\frac{t^5}{5}+2\frac{t^7}{7}-\frac{t^9}{9}=-\frac{1}{5}\cos^5 x+\frac{2}{7}\cos^7 x-\frac{1}{9}\cos^9 x.$$

Example 7:

Evaluate $\int_0^{\pi/4}\sin^5 x\cos^2 x\,dx$.

Solution:

Here the power of sin x being odd, we put cos x = t.

$$\therefore\ \int_0^{\pi/4}\sin^5 x\cos^2 x\,dx=\int_0^{\pi/4}\sin^4 x\cos^2 x.\sin x\,dx$$

$$=\int_0^{\pi/4}(1-\cos^2 x)^2\cos^2 x\ \sin x\,dx=-\int_1^{1/\sqrt{2}}(1-t^2)^2t^2dt,$$

[putting cos x = t so that – sin x dx = dt.]

Also t = 1,

when x = 0

and $t = \frac{1}{\sqrt{2}}$

when $x = \frac{\pi}{4}\Big]$

$$= -\int_1^{1/\sqrt{2}} \left(1 - 2t^2 + t^4\right)t^2 dt = -\int_1^{1/\sqrt{2}} \left(t^2 - 2t^4 + t^6\right)dt$$

$$= -\left[\frac{t^3}{3} - \frac{2t^5}{5} + \frac{t^7}{7}\right]_1^{1/\sqrt{2}}$$

$$= -\left[\left(\frac{1}{6\sqrt{2}} - \frac{1}{10\sqrt{2}} + \frac{1}{56\sqrt{2}}\right) - \left(\frac{1}{3} - \frac{2}{5} + \frac{1}{7}\right)\right]$$

$$= -\left[\frac{71}{840\sqrt{2}} - \frac{8}{105}\right] = \frac{128 - 71\sqrt{2}}{1680}.$$

Example 7:

Evaluate $\int \frac{\cos^5 x}{\sin^2 x} dx.$

Solution:

Let $I = \int \frac{\cos^5 x}{\sin^2 x} dx = \int \frac{\cos^4 x}{\sin^2 x} \cos x\, dx$

$$= \int \frac{\left(1 - \sin^2 x\right)^2}{\sin^2 x} \cos x\, dx.$$

Put sin x = t

so that cos x dx = dt.

Then $I = \int \frac{\left(1 - t^2\right)^2}{t^2} dt = \int \frac{1 - 2t^2 + t^4}{t^2} dt$

$$= \int \left[\frac{1}{t^2} - 2 + t^2\right] dt = -\frac{1}{t} - 2t + \frac{t^3}{3}$$

$$= -\frac{1}{\sin x} - 2\sin x + \frac{\sin^3 x}{3} = -\operatorname{cosec} x - 2\sin x + \frac{1}{3}\sin^3 x.$$

Example 8:

Evaluate $\int \sin^5 x \cos^3 x\, dx.$

Solution:

Here both the powers of cos x and sin x are odd; so we can put either sin x = t or cos x = t.

Let us put sin x = t

so that cos x dx = dt.

$$\therefore \int \sin^5 x \cos^3 x\, dx = \int \sin^5 x \cos^2 x \cos x\, dx$$

$$= \int t^5\left(1-t^2\right)dt = \int\left(t^5 - t^7\right)dt = \frac{1}{6}t^6 - \frac{1}{8}t^8$$

$$= \frac{1}{6}\sin^6 x - \frac{1}{8}\sin^8 x.$$

Example 9:

Evaluate $\int sec\, x\, tan^3 x\, dx$.

Solution:

We have $\int \sec x \tan^3 x\, dx = \int \frac{\sin^3 x}{\cos^4 x}dx.$

Now the power of sin x being odd,

we put cos x = t

so that –sin x dx = dt.

$$\therefore \text{ the given integral } = -\int \frac{\left(1-t^2\right)dt}{t^4} = -\int\left(\frac{1}{t^4} - \frac{1}{t^2}\right)dt$$

$$= \left(\frac{1}{3}\frac{1}{t^3} - \frac{1}{t}\right) = \frac{1}{3\cos^3 x} - \frac{1}{\cos x} = \frac{1}{3}\sec^3 x - \sec x.$$

Example 10:

Evaluate $\int sin^3 x\, cos\, 2x\, dx$.

Solution:

$$I = \int \sin^3 x \cos 2x\, dx = \int \sin^2 x\left(2\cos^2 x - 1\right)\sin x\, dx$$

$$= \int\left(1-\cos^2 x\right)\left(2\cos^2 x - 1\right)\sin x\, dx.$$

Put cos x = t,

so that – sin x dx = dt. Then

$$I = -\int\left(1-t^2\right)\left(2t^2-1\right)dt = -\int\left(3t^2 - 2t^4 - 1\right)dt$$

$$= -\left[t^3 - \frac{2}{3}t^5 - t\right] = -\cos^3 x + \frac{2}{5}\cos^5 x + \cos x.$$

Example 11:

Integrate $\dfrac{1}{\left(\sin^3 x \cos^5 x\right)}.$

Solution:

Here the integrand is $\sin^{-3} x \cos^{-5} x$. It is of the type $\sin^m \cos^n x$, where $m + n = -3 - 5 = -8$ *i.e.* –ive even integer.

$$\therefore \quad I = \int \frac{dx}{\sin^3 x \cos^5 x} = \int \frac{dx}{\left(\sin^3 x/\cos^3 x\right) \cos^3 x . \cos^5 x}$$

$$= \int \frac{\sec^8 x\, dx}{\tan^3 x} = \int \frac{\sec^6 x . \sec^2 x\, dx}{\tan^3 x} \qquad \textbf{(Note)}$$

$$= \int \frac{\left(1 + \tan^2 x\right)^8 \sec^2 x\, dx}{\tan^3 x}.$$

Now put $\tan x = t$

so that $\sec^2 x\, dx = dt$.

$$\therefore \quad I = \int \frac{\left(1 + t^2\right)^3}{t^3} dt = \int \left(\frac{1}{t^3} + \frac{3}{t} + 3t + t^3\right) dt$$

$$= -\left\{\frac{1}{\left(2t^2\right)}\right\} + 3 \log i + \frac{3}{2}t^2 + \frac{1}{4}t^4$$

$$= -\frac{1}{2}\cot^2 x + 3 \log \tan x + \frac{3}{2}\tan^2 x + \frac{1}{4}\tan^4 x.$$

Example 12:

Integrate $\dfrac{1}{\left(\sin^4 x \cos^2 x\right)}.$

Solution:

Here $m = -4$, $n = -2$; $m + n = -6$ *i.e.*, an even negative integer.

$$\therefore \quad \int \frac{dx}{\sin^4 x \cos^2 x} = \int \frac{\sec^6 x}{\tan^4 x} dx = \int \frac{\left(1 + \tan^2 x\right)^2 \sec^2 x\, dx}{\tan^4 x}$$

$$= \int \frac{\left(1 + t^2\right)^2 dt}{t^4} \quad \text{putting } \tan x = t \text{ and } \sec^2 x\, dx = dt$$

$$=\int\left(\frac{1}{t^4}+\frac{2}{t^2}+1\right)dt=-\frac{1}{3t^3}-\frac{2}{t}+t$$

$$=-\frac{1}{3}\cot^3 x-2\cot x+\tan x.$$

Example 13:

Integrate $\dfrac{1}{\sqrt{(\cos^3 x\ \sin^5 x)}}$.

Solution:

Here the integrand is of the type $\cos^m x\ \sin^n x$.

We have $m=-\frac{3}{2}$, $n=-\frac{5}{2}$,

$m+n=-4$ *i.e.*, an even negative integer.

$$\therefore\ \int\frac{dx}{\sqrt{(\cos^3 x\sin^5 x)}}=\int\frac{dx}{\cos^{3/2}\sin^{5/2}x}$$

$$=\int\frac{dx}{\cos^{3/2}x\left(\sin^{5/2}/\cos^{5/2}x\right).\cos^{5/2}x} \qquad \textbf{(Note)}$$

$$=\int\frac{dx}{\cos^4 x\tan^{5/2}x}=\int\frac{\sec^4 x}{\tan^{5/2}x}dx=\int\frac{\sec^2 x}{\tan^{5/2}x}\sec^2 x\,dx$$

$$=\int\frac{\left(1+\tan^2 x\right)}{\tan^{5/2}x}\sec^2 x\,dx.$$

$$=\int\frac{\left(1+t^2\right)}{t^{5/2}}dt, \qquad \text{putting } \tan x=t \text{ and } \sec^2 x\,dx=dt$$

$$=\int\left(t^{-5/2}+t^{-1/2}\right)dt=-\frac{2}{3}t^{-3/2}+2t^{1/2}$$

$$=-\frac{2}{3}(\tan x)^{-3/2}+2(\tan x)^{1/2}=2\sqrt{(\tan x)}-\frac{2}{3}(\tan x)^{-3/2}.$$

Example 14:

Integrate $\sqrt{(\tan x)\sec x\ \operatorname{cosec} x}$.

Solution:

We have $\int\sqrt{(\tan x)\sec x\ \operatorname{cosec} x\,dx}$

$$= \int \sqrt{\left(\frac{\sin x}{\cos x}\right)} \cdot \frac{1}{\cos x \sin x} dx = \int \frac{dx}{\cos^{3/2} \sin^{1/2} x}$$

$$= \int \frac{dx}{\cos^{3/2} x\left(\sin^{1/2} x/\cos^{1/2} x\right)\cos^{1/2} x}$$

$$= \int \frac{dx}{\cos^2 x.\tan^{1/2} x} = \int \frac{\sec^2 x\, dx}{\tan^{1/2} x}$$

$$= \int \left(\tan^{-1/2} x\right)\sec^2 x\, dx = \int \frac{\tan^{1/2} x}{(1/2)}, \quad \text{(by power formula)}$$

$$= 2\sqrt{(\tan x)}.$$

Example 15:

Integrate $sin^2\ x\ cos^4\ x$.

Solution:

We have $\sin^2 x \cos^4 x = (\sin x \cos x)^2 \cos^2 x$

$$= \left(\frac{1}{2}\sin 2x\right)^2 \left\{\frac{1}{2}(1+\cos 2x)\right\}$$

$$= \frac{1}{8}\sin^2 x(1+\cos 2x) = \frac{1}{8}\left\{\frac{1}{2}(1-\cos 4x)\right\}(1+\cos 2x)$$

$$= \frac{1}{16}(1+\cos 2x - \cos 4x - \cos 4x \cos 2x)$$

$$= \frac{1}{16}\left[1+\cos 2x - \cos 4x - \frac{1}{2}(\cos 6x + \cos 2x)\right]$$

$$= \frac{1}{16}\left[1+\frac{1}{2}\cos 2x - \cos 4x - \frac{1}{2}\cos 6x\right]$$

$$= \frac{1}{32}(2+\cos 2x - 2\cos 4x - \cos 6x).$$

$$\therefore \int \sin^2 x \cos^4 x\, dx = \frac{1}{32}\int [2+\cos 2x - 2\cos 4x - \cos 6x]dx$$

$$= \frac{1}{32}\left[2x + \frac{1}{2}\sin 2x - \frac{1}{2}\sin 4x - \frac{1}{6}\sin 6x\right].$$

Example 16:

Evaluate $\int \frac{x^4 dx}{(1+x^2)^2}$.

Solution:

The given integral $I = \int \frac{(x^4 - 1) + 1}{(1 + x^2)^2} dx$

$$= \int \frac{(x^2 - 1)(x^2 + 1)}{(1 + x^2)^2} dx + \int \frac{1}{(1 + x^2)^2} dx$$

$$= \int \frac{x^2 - 1}{x^2 + 1} dx + \int \frac{1}{(1 + x^2)^2} dx$$

$$= \int \frac{(x^2 + 1) - 2}{x^2 + 1} dx + \int \frac{1}{(1 + x^2)^2} dx$$

$$= \int dx - 2\int \frac{1}{1 + x^2} dx + \int \frac{1}{(1 + x^2)^2} dx$$

$$= x - 2\tan^{-1} x + \int \frac{1}{(1 + x^2)^2} dx.$$

Now put $x = \tan\theta$

so that $dx = \sec^2\theta\, d\theta$.

Then $I = x - 2\tan^{-1} x + \int \frac{\sec^2\theta\, d\theta}{\sec^4\theta}$

$$= x - 2\tan^{-1} x + \int \cos^2\theta\, d\theta$$

$$= x - 2\tan^{-1} x + \frac{1}{2}\int (1 + \cos 2\theta)\, d\theta$$

$$= x - 2\tan^{-1} x + \frac{1}{2}\left(\theta + \frac{1}{2}\sin 2\theta\right)$$

$$= x - 2\tan^{-1} x + \frac{1}{2}\theta + \frac{1}{4}\cdot\frac{2\tan\theta}{1 + \tan^2\theta}$$

$$= x - 2\tan^{-1} x + \frac{1}{2}\tan^{-1} x + \frac{1}{2}\frac{x}{1 + x^2}$$

$$= x - \frac{3}{2}\tan^{-1} x + \frac{1}{2}\frac{x}{1 + x^2}.$$

Example 17:

Evaluate $\int x^{-n} \log x\, dx.$

Solution:

Let $I=\int x^{-n}\log x\,dx$.

Integrating by parts taking x^{-n} as the second function and log x as the first function, we get

$$I=\frac{x^{-n+1}}{(-n+1)}\log x-\int\frac{x^{-n+1}}{(-n+1)}\frac{1}{x}dx$$

$$=\frac{1}{(1-n)x^{n-1}}\log x-\frac{1}{(1-n)}\int x^{-n}dx$$

$$=\frac{1}{(1-n)x^{n-1}}\log x-\frac{1}{(1-n)}\frac{x^{-n+1}}{(-n+1)}$$

$$=\frac{\log x}{(1-n)x^{n-1}}-\frac{1}{(1-n)^2x^{n-1}}=\frac{1}{(1-n)x^{n-1}}\left[\log x-\frac{1}{(1-n)}\right].$$

Example 18:

Evaluate $\int x^3(\log x)^2\,dx$.

Solution:

Let $I=\int x^3(\log x)^2\,dx$.

Integrating by parts taking x^3 as the second function and $(\log x)^2$ as the first function, we get

$$I=\frac{1}{4}x^4(\log x)^2-\int\frac{1}{4}x^4.2(\log x).\frac{1}{x}dx$$

$$=\frac{1}{4}x^4(\log x)^2-\frac{1}{2}\int x^3\log x\,dx.$$

Again integrating by parts taking x^3 as the second function, we get

$$I=\frac{1}{4}x^4(\log x)^2-\frac{1}{2}\left[\frac{1}{4}x^4\log x-\int\frac{1}{4}\cdot x^4\cdot\frac{1}{x}dx\right]$$

$$=\frac{1}{4}x^4(\log x)^2-\frac{1}{8}x^4\log x+\frac{1}{8}\int x^3dx$$

$$=\frac{1}{4}x^4(\log x)^2-\frac{1}{8}x^4\log x+\frac{1}{8}\cdot\frac{1}{4}x^4$$

$$=\frac{1}{32}x^4\left[8(\log x)^2-4\log x+1\right].$$

Example 19:

Evaluate $\int \frac{\cosh\theta d\theta}{\sinh\theta + \cosh\theta}$.

Solution:

The given integral

$$I = \int \frac{\frac{(e^{\theta}+e^{-\theta})}{2}}{\frac{1}{2}(e^{\theta}-e^{-\theta})+\frac{1}{2}(e^{\theta}+e^{-\theta})} d\theta$$

$$= \int \frac{(e^{\theta}+e^{-\theta})/2}{e^{\theta}} d\theta = \frac{1}{2}\int (1+e^{-2\theta}) d\theta$$

$$= \frac{1}{2}\left[\theta + \frac{e^{-2\theta}}{-2}\right] = \frac{1}{2}\left[\theta - \frac{1}{2e^{2\theta}}\right] = \frac{2\theta e^{2\theta}-1}{4e^{2\theta}}.$$

Example 20:

Show that $\int_0^{\infty} \frac{dx}{\left[x+\sqrt{(1+x^2)}\right]^n} = \frac{n}{n^2-1}$.

Solution:

Let $I = \int_0^{\infty} \frac{dx}{\left[x+\sqrt{(1+x^2)}\right]^n}$.

Put x = sinh t

so that dx = cosh t dt.

Also when x = 0, sinh t = 0

i.e., t = 0 and when x = ∞,

sin t = ∞, *i.e.*, t = ∞.

$$\therefore \ I = \int_0^{\infty} \frac{\cosh t\, dt}{(\sinh t + \cosh t)^n} = \int_0^{\infty} \frac{\frac{1}{2}(e^t + e^{-t})}{e^{nt}} dt,$$

[∵ sinh t + cosh t = e^t]

$$= \frac{1}{2}\int_0^{\infty}\left[e^{-(n-1)t} + e^{-(n+1)t}\right] dt$$

$$= \frac{1}{2}\left[\frac{e^{-(n-1)t}}{-(n-1)} + \frac{e^{-(n+1)t}}{-(n+1)}\right]_0^{\infty} = \frac{1}{2}\left[\frac{1}{n-1}+\frac{1}{n+1}\right] = \frac{n}{n^2-1}.$$

Example 21:

Evaluate $\int \frac{dx}{(a^2 + b^2x^2)^{3/2}}$.

Solution:

Put bx = a tan θ,

so that b dx = a sec² θ dθ.

Then the given integral

$$I = \int \frac{1}{(a^2 + a^2\tan^2\theta)^{3/2}} \frac{a}{b} \sec^2\theta\, d\theta = \frac{1}{a^2 b}\int \cos\theta\, d\theta$$

$$= \frac{1}{a^2 b}\sin\theta.$$

Now $\tan\theta = \frac{bx}{a}$ gives $\sin\theta = \frac{bx}{\sqrt{(a^2 + b^2x^2)}}$.

$$\therefore\ I = \frac{1}{a^2 b}\frac{bx}{\sqrt{(a^2 + b^2x^2)}} = \frac{x}{a^2\sqrt{(a^2 + b^2x^2)}}.$$

Example 22:

Evaluate $\int (ax^2 + c)^{-3/2}\, dx$.

Solution:

The given integral $I = \int \frac{dx}{(ax^2 + c)^{3/2}}$.

Put $\sqrt{ax} = \sqrt{c}\tan\theta$ and do your self. **Ans.** $\frac{x}{c\sqrt{(ax^2 + c)}}$.

Example 23:

Evaluate $\int \frac{\sec x\ \text{cosec}\, x}{\log\tan x}\, dx$.

Solution:

Put log tan x = t,

so that $\frac{1}{\tan x}\sec^2 x\, dx = dt$

i.e., sec x cosec x dx = dt.

Then the given integral

$$I = \int \frac{dt}{t} = \log t = \log(\log \tan x).$$

Example 24:

Evaluate the following integrals:

(i) $\int \frac{x\,dx}{1+\sin x}$

(ii) $\int \frac{x}{\operatorname{cosec} x + 1} dx.$

Solution:

(i) The given integral

$$I = \int \frac{x\,dx}{1+\sin x}$$

$$= \int \frac{x(1-\sin x)}{(1+\sin x)(1-\sin x)} dx = \int \frac{x(1-\sin x)}{1-\sin^2 x} dx$$

$$= \int x\left[\frac{1-\sin x}{\cos^2 x}\right] dx$$

$$= \int x\left(\sec^2 x - \sec x \tan x\right) dx.$$

Integrating by parts taking x as the first function, we have

$I = x(\tan x - \sec x) - \int (\tan x - \sec x) dx$

$= x(\tan x - \sec x) - \int \tan x\,dx + \int \sec x\,dx$

$= x(\tan x - \sec x) + \log \cos x + \log(\sec x + \tan x)$

$= x(\tan x - \sec x) + \log\{\cos x(\sec x + \tan x)\}$

$= x(\tan x - \sec x) + \log(1 + \sin x).$

(ii) The given integral

$$I = \int \frac{x \sin x}{1+\sin x} dx = \int \frac{x(1+\sin x - 1)}{1+\sin x} dx$$

$$= \int x\left[\frac{1+\sin x}{1+\sin x} - \frac{1}{1+\sin x}\right] dx$$

$$= \int x\left[1 - \frac{1}{1+\sin x}\right] dx$$

$$= \int x\,dx - \int \frac{x}{1+\sin x}dx.$$

Now proceeding as in part (i), we get

$$I = \frac{1}{2}x^2 - x(\tan x - \sec x) - \log(1+\sin x).$$

Example 25:

Evaluate the following integrals:

(i) $\int \frac{x\,dx}{1+\cos x}$

(ii) $\int \frac{x\,dx}{\sec x + 1}$.

Solution:

(i) The given integral

$$I = \int \frac{x\,dx}{2\cos^2 \frac{1}{2}x} = \frac{1}{2}\int x \sec^2 \frac{1}{2}x\,dx.$$

Integrating by parts taking $\sec^2 \frac{1}{2}x$ as the second function, we have

$$I = \frac{1}{2}\left[x.2\tan\frac{1}{2}x - \int 1.2\tan\frac{1}{2}x\,dx\right]$$

$$= x\tan\frac{1}{2}x - \int \tan\frac{1}{2}x\,dx$$

$$= x\tan\frac{1}{2}x - 2\log\sec\frac{1}{2}x, \quad \left[\because \int \tan\frac{1}{2}x\,dx = 2\log\sec\frac{1}{2}x\right].$$

(ii) The given integral

$$I = \int \frac{x\cos x\,dx}{1+\cos x} = \int \frac{x(1+\cos x - 1)}{1+\cos x}dx$$

$$= \int x\left[1 - \frac{1}{1+\cos x}\right]dx$$

$$= \int x\,dx - \int \frac{x\,dx}{1+\cos x}.$$

Now proceeding as in part (i), we get

$$I = \frac{1}{2}x^2 - x\tan\frac{1}{2}x + 2\log\sec\frac{1}{2}x.$$

Example 26:

Evaluate the following integrals:

(i) $\int \cos\left(a\log\frac{x}{b}\right)dx$

(ii) $\int \frac{\sin(\log x)}{x^3}dx.$

Solution:

(i) Let $I = \int \cos\left(a\log\frac{x}{b}\right)dx.$

Put $\log\left(\frac{x}{b}\right) = t,$

i.e., $\frac{x}{b} = e^t$

i.e., $x = be^t.$

Then $dx = be^t\ dt$

$\therefore\ I = \int be^t \cos at\, dt = b\int e^t \cos at\, dt.$

But $\int e^{ax}\cos bx\, dx = \frac{e^{ax}}{\sqrt{(a^2+b^2)}}\cos\left(bx - \tan^{-1}\frac{b}{a}\right).$

$$\therefore\ I = b\frac{e^t}{\sqrt{(1+a^2)}}\cos\left\{at - \tan^{-1}\frac{a}{1}\right\}$$

$$= b\cdot\frac{x}{b}\cdot\frac{1}{\sqrt{(1+a^2)}}\cos\left\{a\log\frac{x}{b} - \tan^{-1}a\right\}$$

$$= \frac{x}{\sqrt{(1+a^2)}}\cos\left\{a\log\frac{x}{b} - \tan^{-1}a\right\}.$$

(ii) Let $I = \int \frac{\sin(\log x)}{x^3}dx.$

Put $\log x = t$ *i.e.,* $x = e^t.$

Then $dx = e^t\ dt.$

$\therefore\ I = \int \frac{\sin t}{e^{3t}}e^t dt = \int e^{-2t}\sin t\, dt.$

But $\int e^{ax}\sin bx\, dx = \frac{e^{ax}}{a^2+b^2}(a\sin bx - b\cos bx).$

$$\therefore\ I = \frac{e^{-2t}}{(-2)^2 + (1)^2}(-2\sin t - 1.\cos t)$$

$$= \frac{1}{5}e^{-2t}(-2\sin t - \cos t)$$

$$= \frac{1}{5}\cdot\frac{1}{x^2}(-2\sin \log x - \cos \log x)$$

$$= -\frac{1}{5x^2}\ (2 \sin \log x + \cos \log x).$$

Example 27:

Evaluate the following integrals:

(i) $\displaystyle\int \frac{x\, e^{\sin^{-1} x}}{\sqrt{(1-x^2)}} dx$,

(ii) $\displaystyle\int e^{\sin^{-1} x} dx$

Solution:

(i) Let $I = \displaystyle\int \frac{x e^{\sin^{-1} x}}{\sqrt{(1-x^2)}} dx.$

Put $\sin^{-1} x = t$,

so that $\left[\dfrac{1}{\sqrt{(1-x^2)}}\right] dx = dt.$

Also $x = \sin t$.

$\therefore\ I = \displaystyle\int \sin t . e^t\, dt.$

But $\displaystyle\int e^{ax} \sin bx\, dx = \frac{e^{ax}}{a^2 + b^2}(a \sin bx - b \cos bx).$

$$\therefore\ I = \frac{e^t}{1+1}(\sin t - \cos t) = \frac{e^t}{2}\left\{\sin t - \sqrt{(1-\sin^2 t)}\right\}$$

$$= \frac{1}{2}e^{\sin^{-1} x}\left\{x - \sqrt{(1-x^2)}\right\}.$$

(ii) Let $I = \int e^{\sin^{-1}x} dx$.

Put $\sin^{-1} x = t$ *i.e.*, $x = \sin t$,

so that $dx = \cos t \, dt$.

Then $I = \int e^t \cos t \, dt$

$= e^t \cos t - \int e^t(-\sin t)dt$,

integrating by parts taking e^t as the second function

$= e^t \cos t + \int e^t \sin t \, dt$

$= e^t \cos t + e^t \sin t - \int e^t \cos t \, dt$,

again integrating by parts taking e^t as the second function

$= e^t (\cos t + \sin t) - I$.

$\therefore 2I = e^t (\cos t + \sin t)$

or $I = \frac{1}{2} e^t \left\{\sin t + \sqrt{(1 - \sin^2 t)}\right\}$

$= \frac{1}{2} e^{\sin^{-1}x} \left\{x + \sqrt{(1 - x^2)}\right\}$.

Example 28:

Evaluate the following integrals:

(i) $\int \tan^{-1} \sqrt{x} \, dx$,

(ii) $\int \frac{\cos^{-1}x}{x^3} dx$,

(iii) $\int \frac{\tan^{-1}x}{(1+x)^2} dx$.

Solution:

(i) Let $I = \int \tan^{-1} \sqrt{x} \, dx$.

Put $\sqrt{x} = \tan t$

i.e., $x = \tan^2 t$,

so that $dx = 2 \tan t \sec^2 t \, dt$.

Then $I = \int (\tan^{-1} \tan t) 2 \tan t \sec^2 t \, dt$

$$= \int t\left(2\tan t \sec^2 t\right)dt.$$

Integrating by parts taking 2 tan t $\sec^2$ t as the second function, we have

$$I = t\tan^2 t - \int 1.\tan^2 t\,dt = t\tan^2 t - \int\left(\sec^2 t - 1\right)dt$$

$$= t\tan^2 t - \int \sec^2 t\,dt + \int dt$$

$$= t\tan^2 t - \tan t + t = t\left(\tan^2 t + 1\right) - \tan t$$

$$= (x+1)\tan^{-1}\sqrt{x} - \sqrt{x}.$$

(ii) Let $I = \int \frac{\cos^{-1} x}{x^3} dx.$

Put $\cos^{-1} x = t$

i.e., x = cos t,

so that dx = – sin t dt.

Then $I = \int \frac{t}{\cos^3 t}(-\sin t)dt$

$$= -\int t \tan t \sec^2 t\,dt.$$

Integrating by parts taking tan t $\sec^2$ t as the second function, we have

$$I = -t\left(\frac{1}{2}\tan^2 t\right) + \int 1\cdot\frac{1}{2}\tan^2 t\,dt$$

$$= -\frac{1}{2}t\tan^2 t + \frac{1}{2}\int\left(\sec^2 t - 1\right)dt$$

$$= -\frac{1}{2}t\tan^2 t + \frac{1}{2}\tan t - \frac{1}{2}t$$

$$= -\frac{1}{2}t\left(1 + \tan^2 t\right) + \frac{1}{2}\tan t = -\frac{1}{2}t\sec^2 t + \frac{1}{2}\tan t.$$

But if cos t = x, then $\tan t = \frac{x}{\sqrt{(1-x^2)}}$.

$$\therefore \quad I = -\frac{1}{2}\frac{\cos^{-1} x}{x^2} + \frac{1}{2}\frac{x}{\sqrt{(1-x^2)}}.$$

(iii) Let $I = \int \frac{\tan^{-1} x}{(1+x)^2} dx.$

The integral of $\frac{1}{(1+x)^2}$ is $-\frac{1}{(1+x)}$. So integrating by parts taking $\frac{1}{(1+x)^2}$ as the second function, we have

$$I = \left(\tan^{-1}x\right)\left(\frac{-1}{1+x}\right) - \int \frac{1}{\left(1+x^2\right)} \cdot \frac{-1}{(1+x)} dx$$

$$= -\frac{\tan^{-1}x}{1+x} + \int \frac{1}{(1+x)\left(1+x^2\right)} dx.$$

Now let $\frac{1}{(1+x)\left(1+x^2\right)} = \frac{A}{1+x} + \frac{Bx+C}{1+x^2}$.

Then $1 \equiv A\,(1 + x^2) + (Bx + C)\,(1 + x)$ ...(1)

Putting $x = -1$ on both sides of (1), we have

$1 = 2A$ *i.e.*, $A = \frac{1}{2}$.

Putting $A = \frac{1}{2}$ in (1), we have

$1 \equiv \frac{1}{2}\left(1+x^2\right) + (Bx+C)\,(1+x)$ ...(2)

Equating the coefficients of x^2 and constant terms on both sides of (2), we get

$$0 = \frac{1}{2} + B,\ 1 = \frac{1}{2} + C.$$

These give $B = -\frac{1}{2}$ and $C = \frac{1}{2}$.

$$\therefore\quad I = -\frac{\tan^{-1}x}{1+x} + \int \left[\frac{\frac{1}{2}}{1+x} + \frac{-\frac{1}{2}x + \frac{1}{2}}{1+x^2}\right] dx$$

$$= -\frac{\tan^{-1}x}{1+x} + \frac{1}{2}\int \frac{1}{1+x} dx - \frac{1}{2}\int \frac{\frac{1}{2}\cdot(2x)}{1+x^2} dx + \frac{1}{2}\int \frac{1}{1+x^2} dx$$

$$= -\frac{\tan^{-1}x}{1+x} + \frac{1}{2}\log(1+x) - \frac{1}{4}\log\left(1+x^2\right) + \frac{1}{2}\tan^{-1}x$$

$$= \left(\frac{1}{2} - \frac{1}{1+x}\right)\tan^{-1}x + \frac{1}{4}\log\frac{(1+x)^2}{\left(1+x^2\right)}.$$

Example 29:

Evaluate the following integrals:

(i) $\int \frac{\sin x}{\sin(x-a)} dx$

(ii) $\int \frac{dx}{\sin(x-\alpha)\sin(x-\beta)}$.

Solution:

(i) Let $I = \int \frac{\sin x}{\sin(x-\alpha)} dx$.

Put $x - \alpha = t$,

so that $dx = dt$.

Then $I = \int \frac{\sin(\alpha + t)}{\sin t} dt = \int \frac{\sin\alpha \cos t + \cos\alpha \sin t}{\sin t} dt$

$= \sin\alpha \int \frac{\cos t}{\sin t} dt + \cos\alpha \int dt$

$= \sin\alpha \log \sin t + t \cos\alpha$

$= \sin\alpha \log \sin(x-\alpha) + (x-\alpha)\cos\alpha$

$= x\cos\alpha + \sin\alpha \log \sin(x-\alpha)$,

because the constant term $-\alpha\cos\alpha$ may be added to the constant of integration c.

(ii) Let $I = \int \frac{dx}{\sin(x-\alpha)\sin(x-\beta)}$

$= \int \frac{[\sin\{(x-\beta)-(x-\alpha)\}].[1/\sin(\alpha-\beta)]}{\sin(x-\alpha)\sin(x-\beta)}$ **(Note)**

$= \frac{1}{\sin(\alpha-\beta)} \int \frac{\sin(x-\beta)\cos(x-\alpha) - \cos(x-\beta)\sin(x-\alpha)}{\sin(x-\alpha)\sin(x-\beta)} dx$

$= \operatorname{cosec}(\alpha-\beta)\left[\int \frac{\cos(x-\alpha)}{\sin(x-\alpha)} dx - \int \frac{\cos(x-\beta)}{\sin(x-\beta)} dx\right]$

$= \operatorname{cosec}(\alpha-\beta)[\log \sin(x-\alpha) - \log \sin(x-\beta)]$

$= \operatorname{cosec}(\alpha-\beta) \log \frac{\sin(x-\alpha)}{\sin(x-\beta)}$.

Example 30:

Evaluate the following integrals:

(i) $\int \sqrt{\left\{\frac{\sin(x-\alpha)}{\sin(x+\alpha)}\right\}}\,dx$

(ii) $\int \frac{dx}{\sqrt{\{\sin^3 x \sin(x+\alpha)\}}}$.

Solution:

(i) Let $I = \int \sqrt{\left\{\frac{\sin(x-\alpha)}{\sin(x+\alpha)}\right\}}\,dx$

$$= \int \frac{\sin(x-\alpha)}{\sqrt{\{\sin(x+\alpha)+\sin(x-\alpha)\}}},$$

multiplying the numerator and denominator by $\sqrt{\{\sin(x-\alpha)\}}$

$$= \int \frac{\sin x \cos\alpha - \cos x \sin\alpha}{\sqrt{\{\sin(x+\alpha)\sin(x-\alpha)\}}}dx$$

$$= \int \frac{\sin x \cos\alpha}{\sqrt{\{\sin(x+\alpha)\sin(x-\alpha)\}}}dx - \int \frac{\cos x \sin\alpha}{\sqrt{\{\sin(x+\alpha)\sin(x-\alpha)\}}}dx$$

$$= \int \frac{\sin x \cos\alpha}{\sqrt{(\cos^2\alpha - \cos^2 x)}}dx - \int \frac{\cos x \sin\alpha}{\sqrt{(\sin^2 x - \sin^2\alpha)}}dx.$$

[$\because$ $\sin(A+B)\sin(A-B) = \sin^2 A - \sin^2 B = \cos^2 B - \cos^2 A$]

In the first integral put $\cos x = t$ so that $-\sin x\, dx = dt$ and in the second integral put $\sin x = z$ so that $\cos x\, dx = dz$.

Then $I = \cos\alpha \int \frac{-dt}{\sqrt{(\cos^2\alpha - t^2)}} - \sin\alpha \int \frac{dz}{\sqrt{(z^2 - \sin^2\alpha)}}$

$$= \cos\alpha \cos^{-1}\left(\frac{t}{\cos\alpha}\right) - \sin\alpha \cosh^{-1}\left(\frac{z}{\sin\alpha}\right)$$

$= \cos\alpha \cos^{-1}(\cos x \text{ sex } \alpha) - \sin\alpha \cosh^{-1}(\sin x \operatorname{cosec}\alpha)$.

(ii) The given integral

$$I = \int \frac{dx}{\sqrt{\{\sin^3 x(\sin x \cos\alpha + \cos x \sin\alpha)\}}}$$

$$= \int \frac{dx}{\sqrt{\{\sin^4 x(\cos\alpha + \cot x \sin\alpha)\}}}$$

$$= \int \frac{dx}{\sin^2 x\sqrt{(\cos\alpha + \cot x \sin\alpha)}} = \int \frac{\operatorname{cosec}^2 x\, dx}{\sqrt{(\cos\alpha + \cot x \sin\alpha)}}$$

$$= -\frac{1}{\sin\alpha}\int(\cos\alpha + \cot x \sin\alpha)^{-1/2}\left(-\sin\alpha \ \text{cosec}^2 x\right)dx,$$

adjusting suitably to apply the power formula

$$= -\frac{1}{\sin\alpha}\frac{(\cos\alpha + \cot x \sin\alpha)^{1/2}}{1/2}, \text{ power formula}$$

$$= -2\,\text{cosec}\,\alpha\left\{\cos\alpha + \left(\frac{\cos x}{\sin x}\right)\sin\alpha\right\}^{1/2}$$

$$= -2\,\text{cosec}\,\alpha\sqrt{\left\{\frac{\sin(x+\alpha)}{\sin x}\right\}}.$$

Example 31:

Evaluate the following integrals:

(i) $\int\frac{(\sin\theta - \cos\theta)}{\sqrt{(\sin 2\theta)}}d\theta$

(ii) $\int\frac{(\sin x - \cos x)dx}{(\sin x + \cos x)\sqrt{(\sin x\cos x + \sin^2 x\cos^2 x)}}.$

Solution:

(i) Let $I = \int\frac{(\sin\theta - \cos\theta)}{\sqrt{(\sin 2\theta)}}d\theta$

$$= \int\frac{(\sin\theta - \cos\theta)}{\sqrt{(1 + \sin 2\theta - 1)}}d\theta \qquad \textbf{(Note)}$$

$$= \int\frac{(\sin\theta - \cos\theta)d\theta}{\sqrt{\left\{\left(\sin^2\theta + \cos^2\theta + 2\sin\theta\cos\theta\right) - 1\right\}}}$$

$$= \int\frac{(\sin\theta - \cos\theta)d\theta}{\sqrt{\left\{(\sin\theta + \cos\theta)^2 - 1\right\}}}.$$

Put $\sin\theta + \cos\theta = t$,

so that $(\cos\theta - \sin\theta)\,d\theta = dt$.

Then $I = -\int\frac{dt}{\sqrt{(t^2 - 1)}} = -\cosh^{-1} t = -\cosh^{-1}(\sin\theta + \cos\theta).$

(ii) Let $I = \int \frac{(\sin x - \cos x)dx}{(\sin x + \cos x)\sqrt{(\sin x \cos x + \sin^2 x \cos^2 x)}}$

$$= \int \frac{(\sin x - \cos x)(\sin x + \cos x)dx}{(\sin x + \cos x)^2 \cdot \frac{1}{2}\sqrt{(4\sin x \cos x + 4\sin^2 x \cos^2 x)}} \quad \textbf{(Note)}$$

$$= \int \frac{-2(\cos^2 x - \sin^2 x)dx}{(\sin^2 x + \cos^2 x + 2\sin x \cos x)\sqrt{\{(1 + 4\sin x \cos x + 4\sin^2 x \cos^2 x) - 1\}}}$$

$$= \int \frac{-2\cos 2x\, dx}{(1 + \sin 2x)\sqrt{\{(1 + \sin 2x)^2 - 1\}}}.$$

Put $1 + \sin 2x = t$,

so that $2 \cos 2x\, dx = dt$.

Then $I = -\int \frac{dt}{t\sqrt{(t^2 - 1)}} = \operatorname{cosec}^{-1} t = \operatorname{cosec}^{-1}(1 + \sin 2x)$.

Example 32:

Evaluate the integrals:

(i) $\int_0^{\pi/4} \frac{\sin 2\theta}{\sin^4\theta + \cos^4\theta} d\theta$

(ii) $\int \sqrt{\left\{\frac{1 - \cos\theta}{\cos\theta(1 + \cos\theta)(2 + \cos\theta)}\right\}} d\theta.$

Solution:

(i) Let

$$I = \int_0^{\pi/4} \frac{\sin 2\theta\, d\theta}{\sin^4\theta + \cos^4\theta} = \int_0^{\pi/4} \frac{2\sin\theta \cos\theta}{\sin^4\theta + \cos^4\theta} d\theta$$

$$= \int_0^{\pi/4} \frac{2\tan\theta \sec^2\theta\, d\theta}{1 + \tan^4\theta},$$

dividing the numerator and denominator by $\cos^4\theta$.

Put $\tan^2\theta = t$,

so that $2 \tan\theta \sec^2\theta\, d\theta = dt$.

When $\theta = 0$,

$t = \tan^2 0 = 0$

and when $\theta = \frac{\pi}{4}$,

$$t = \tan^2 \frac{1}{4}\pi = 1.$$

$$\therefore \quad I = \int_0^1 \frac{dt}{1+t^2} = \left[\tan^{-1} t\right]_0^1$$

$$= \tan^{-1} 1 - \tan^{-1} 0 = \frac{\pi}{4} - 0 = \frac{\pi}{4}.$$

(ii) Multiplying the numerator and denominator by $\sqrt{(1+\cos\theta)}$, the given integral

$$I = \int \frac{\sqrt{\{(1-\cos\theta)(1+\cos\theta)\}}}{(1+\cos\theta)\sqrt{\{\cos\theta(2+\cos\theta)\}}} d\theta$$

$$= \int \frac{\sin\theta \, d\theta}{(1+\cos\theta)\sqrt{(\cos^2\theta + 2\cos\theta)}}$$

$$= \int \frac{\sin\theta \, d\theta}{(1+\cos\theta)\sqrt{\{(1+\cos\theta)^2 - 1\}}}. \qquad \textbf{(Note)}$$

Now put $1 + \cos\theta = t$,

so that $-\sin\theta \, d\theta = dt$.

Then $I = -\int \frac{dt}{t\sqrt{(t^2-1)}}$

$$= \operatorname{cosec}^{-1} t = \operatorname{cosec}^{-1}(1+\cos\theta) = \operatorname{cosec}^{-1}\left(2\cos^2\frac{\theta}{2}\right).$$

Example 33:

Evaluate $\int_0^{\pi/2} \sin^4 x \cos^2 x \, dx$.

Solution:

Let $y = \cos x + i \sin x$. Then $y^{-1} = \cos x - i \sin x$.

We have $(y + y^{-1}) = 2\cos x$ and $(y - y^{-1}) = 2i \sin x$.

$\therefore$ $(2i \sin x)^4 (2\cos x)^2 = (y - y^{-1})^4 (y + y^{-1})^2$

or $2^6 i^4 \sin^4 x \cos^2 x$

$$= \left(y^4 - 4y^3 \cdot \frac{1}{y} + 6y^2 \cdot \frac{1}{y^2} - 4y \cdot \frac{1}{y^3} + \frac{1}{y^4}\right)\left(y + \frac{1}{y}\right)\left(y + \frac{1}{y}\right)$$

$$=\left[\left(y^4-4y^2+6-\frac{4}{y^2}+\frac{1}{y^4}\right)\left(y+\frac{1}{y}\right)\right]\left(y+\frac{1}{y}\right)$$

$$=\left(y^5-3y^3+2y+\frac{2}{y}-\frac{3}{y^3}+\frac{1}{y^5}\right)\left(y+\frac{1}{y}\right)$$

$$=y^6-2y^4-y^2+4-\frac{1}{y^2}-\frac{2}{y^4}+\frac{1}{y^6}.$$

$$\therefore\ 64\sin^4 x\,.\cos^2 x=\left(y^6+\frac{1}{y^6}\right)-2\left(y^4+\frac{1}{y^4}\right)-\left(y^2+\frac{1}{y^2}\right)+4$$

$$= 2\cos 6x - 2.2\cos 4x - 2\cos 2x + 4.$$

$$\therefore\ \int_0^{\pi/2}\sin^4 x\cos^2 x\,dx$$

$$=\frac{1}{32}\int_0^{\pi/2}[\cos 6x-2\cos 4x-\cos 2x+2]dx$$

$$=\frac{1}{32}\left[\frac{\sin 6x}{6}-\frac{\sin 4x}{4}-\frac{\sin 2x}{2}+2x\right]_0^{\pi/2}=\frac{1}{32}\cdot 2\cdot\frac{\pi}{2}=\frac{\pi}{32}.$$

Example 34:

If m and n are integers, prove that

$\int_0^{\pi}\cos mx\sin nx\,dx=\dfrac{2n}{n^2-m^2}$ *or 0 according as (n – m) is odd or even.*

Solution:

$$\text{The given integral }=\frac{1}{2}\int_0^{\pi}2\cos mx\ \sin nx\,dx$$

$$=\frac{1}{2}\int_0^{\pi}2\sin nx\ \cos mx\,dx$$

$$=\frac{1}{2}\int_0^{\pi}[\sin(m+n)x+\sin(n-m)x]dx,\ \text{by trigonometry}$$

$$=-\frac{1}{2}\left[\frac{\cos(m+n)x}{m+n}+\frac{\cos(n-m)x}{n-m}\right]_0^{\pi},\ \text{if } n-m\neq 0\ \textit{i.e.},\ \text{if } n\neq m$$

$$=-\frac{1}{2}\left[\left\{\frac{(-1)^{m+n}}{m+n}+\frac{(-1)^{n-m}}{n-m}\right\}-\left\{\frac{1}{m+n}+\frac{1}{n-m}\right\}\right],$$

$[\because \cos r\pi = (-1)^r]$

Casé 1: n – m is odd.

When n – m is odd, n + m is also odd because we can write

$$n + m = (n - m) + 2m.$$

$\therefore$ In this case $(-1)^{m+n} = (-1)^{n-m} = -1$.

Hence in this case the given integral

$$= -\frac{1}{2}\left[-\frac{1}{m+n} - \frac{1}{n-m} - \frac{1}{m+n} - \frac{1}{n-m}\right] = \frac{2n}{n^2 - m^2}.$$

Case 2: n – m is even.

If (n – m) is even, then n + m is also ever.

$\therefore (-1)^{n-m} = (-1)^{n+m} = 1.$

Then the given integral

$$= -\frac{1}{2}\left[-\frac{1}{m+n} + \frac{1}{n-m} - \frac{1}{m+n} - \frac{1}{n-m}\right] = 0.$$

Now if n = m, then n – m = 0 which is even. Also in this case the given integral

$$= \frac{1}{2}\int_0^{\pi} 2\sin mx \cos mx\, dx = \frac{1}{2}\int_0^{\pi} \sin 2mx\, dx$$

$$= \frac{1}{2}\left[-\frac{1}{2m}\cos 2mx\right]_0^{\pi}$$

$$= -\frac{1}{4m}[\cos 2m\pi - \cos 0] = -\frac{1}{4m}(1-1) = 0.$$

Hence the result follows.

Example 35:

If m and n are integers, show that

$$\int_0^{\pi} \sin mx.\sin nx\, dx = 0 \text{ if } m \neq n \text{ and } = \frac{\pi}{2} \text{ if } m = n.$$

Solution:

The given integral

$$= \frac{1}{2}\int_0^{\pi} 2\sin mx.\sin nx\, dx$$

$$= \frac{1}{2}\int_0^{\pi} [\cos(m-n)x - \cos(m+n)x]dx$$

$$= \frac{1}{2}\left[\frac{\sin(m-n)x}{m-n} - \frac{\sin(m+n)x}{m+n}\right]_0^{\pi}, \quad \text{if } m - n \neq 0 \text{ i.e., if } m \neq n$$

$$= \frac{1}{2}[(0-0)-(0-0)] = 0, \qquad [\because \sin r\pi = 0].$$

Now when m = n, the given integral

$$= \int_0^{\pi} \sin^2 nx\, dx$$

$$= \frac{1}{2}\int_0^{\pi}(1-\cos 2nx)dx = \frac{1}{2}\left[x - \frac{\sin 2nx}{2n}\right]_0^{\pi}$$

$$= \frac{1}{2}[(\pi-0)-(0-0)] = \frac{1}{2}\pi.$$

Example 36:

Evaluate $\int_0^{\pi/4} \cos 3x \cos 5x\, dx$.

Solution:

The given integral $= \frac{1}{2}\int_0^{\pi/4} 2\cos 3x \cos 5x\, dx$

$$= \frac{1}{2}\int_0^{\pi/4}(\cos 8x + \cos 2x)dx$$

$$= \frac{1}{2}\left[\frac{\sin 8x}{8} + \frac{\sin 2x}{2}\right]_0^{\pi/4}$$

$$= \frac{1}{2}\left[\left\{\frac{\sin 2\pi}{8} + \frac{\sin\frac{1}{2}\pi}{2}\right\} - \left\{\frac{\sin 0}{8} + \frac{\sin 0}{2}\right\}\right]$$

$$= \frac{1}{2}\left[\frac{1}{2}\right] = \frac{1}{4}.$$

Example 37:

Evaluate $\int \cos x \cos 2x \cos 3x\, dx$.

Solution:

The given integral $= \frac{1}{2}\int \cos x.(2\cos 2x \cos 3x)dx$

$$=\frac{1}{2}\int \cos x(\cos 5x+\cos x)\,dx$$

$$=\frac{1}{4}\int 2\cos x\cos 5x\,dx+\frac{1}{4}\int 2\cos^2 x\,dx$$

$$=\frac{1}{4}\int(\cos 6x+\cos 4x)\,dx+\frac{1}{4}\int(1+\cos 2x)\,dx$$

$$=\frac{1}{4}\int \cos 6x\,dx+\frac{1}{4}\int \cos 4x\,dx+\frac{1}{4}\int \cos 2x\,dx+\frac{1}{4}\int dx$$

$$=\frac{1}{4}\cdot\frac{\sin 6x}{6}+\frac{1}{4}\,\frac{\sin 4x}{4}+\frac{1}{4}\,\frac{\sin 2x}{2}+\frac{1}{4}\cdot x$$

$$=\frac{1}{4}\left[\left(\frac{1}{6}\right)\cdot\sin 6x+\frac{1}{4}\sin 4x+\frac{1}{2}\sin 2x+x\right].$$

Example 38:

Integrate $\dfrac{1}{(5+4\cos x)}$.

Solution:

We have $\displaystyle\int\frac{dx}{5+4\cos x}$

$$=\int\frac{dx}{5\left(\cos^2\frac{1}{2}x+\sin^2\frac{1}{2}x\right)+4\left(\cos^2\frac{1}{2}x-\sin^2\frac{1}{2}x\right)}\qquad\textbf{(Note)}$$

$$=\int\frac{dx}{9\cos^2\frac{1}{2}x+\sin^2\frac{1}{2}x}=\int\frac{\sec^2\frac{1}{2}x\,dx}{9+\tan^2\frac{1}{2}x},$$

dividing the numerator and the denominator by $\cos^2\frac{1}{2}x$.

Now putting $\tan\frac{1}{2}x=t$

so that $\frac{1}{2}\sec^2\frac{1}{2}x\,dx=dt$,

the required integral $=2\displaystyle\int\frac{dt}{9+t^2}$

$$=2\cdot\frac{1}{3}\tan^{-1}\left(\frac{t}{3}\right)=\frac{2}{3}\tan^{-1}\left(\frac{1}{3}\tan\frac{1}{2}x\right).$$

Example 39:

Evaluate $\int_0^{\pi/2} \frac{dx}{5+4\cos x}$.

Solution:

Proceeding as in part (a), we get

$\int_0^{\pi/2} \frac{dx}{5+4\cos x} = 2\int_0^1 \frac{dt}{9+t^2}$. [Note that when x = 0, t = tan 0 = 0 and when x = π/2, t = tan 1/4π = 1]

$$= 2\times\frac{1}{3}\left[\tan^{-1}\frac{1}{3}t\right]_0^1 = \frac{2}{3}\left[\tan^{-1}\frac{1}{3} - \tan^{-1}0\right]$$

$$= \frac{2}{3}\left[\left(\tan^{-1}\frac{1}{3}\right) - 0\right] = \frac{2}{3}\tan^{-1}\frac{1}{3}.$$

Example 40:

Evaluate $\int_0^{\pi/2} \frac{dx}{4+5\cos x}$.

Solution:

We have $I = \int_0^{\pi/2} \frac{dx}{4+5\cos x}$

$$= \int_0^{\pi/2} \frac{dx}{4\left(\cos^2\frac{1}{2}x + \sin^2\frac{1}{2}x\right) + 5\left(\cos^2\frac{1}{2}x - \sin^2\frac{1}{2}x\right)} \quad \textbf{(Note)}$$

$$= \int_0^{\pi/2} \frac{dx}{9\cos^2\frac{1}{2}x - \sin^2\frac{1}{2}x} = \int_0^{\pi/2} \frac{\sec^2\frac{1}{2}x\,dx}{9 - \tan^2\frac{1}{2}x}.$$

Now put $\tan\frac{1}{2}x = t$

so that $\frac{1}{2}\sec^2\frac{1}{2}x\,dx = dt$.

Also t = 0 when x = 0

and t = 1 when $x = \frac{\pi}{2}$.

$$\therefore\ I = \int_0^1 \frac{2\,dt}{9-t^2} = 2\cdot\frac{1}{2.3}\left[\log\frac{3+t}{3-t}\right]_0^1$$

$$= \frac{1}{3}\left[\log\frac{3+1}{3-1} - \log\frac{3+0}{3-0}\right] = \frac{1}{3}\log 2.$$

Example 41:

Integrate $\dfrac{1}{(3+2\cos x)}$.

Solution:

We have $\displaystyle\int\frac{dx}{3+2\cos x} = \int\frac{dx}{5\cos^2\frac{1}{2}x + \sin^2\frac{1}{2}x}$

$$= \int\frac{\sec^2\frac{1}{2}x\,dx}{5+\tan^2\frac{1}{2}x} = 2\int\frac{dt}{5+t^2},$$

putting $\tan\frac{1}{2}x = t$

so that $\frac{1}{2}\sec^2\frac{1}{2}x\,dx = dt$

$$= 2\cdot\frac{1}{\sqrt{5}}\tan^{-1}\left(\frac{t}{\sqrt{5}}\right) = \frac{2}{\sqrt{5}}\tan^{-1}\left(\frac{1}{\sqrt{5}}\tan\frac{1}{2}x\right).$$

Example 42:

Integrate $\dfrac{1}{(2+\cos x)}$.

Solution:

$$\int\frac{dx}{2+\cos x} = \int\frac{dx}{3\cos^2\frac{1}{2}x + \sin^2\frac{1}{2}x} = \int\frac{\sec^2\frac{1}{2}x\,dx}{3+\tan^2\frac{1}{2}x} \qquad \textbf{(Note)}$$

$$= 2\int\frac{dt}{3+t^2},$$

putting $\tan\frac{1}{2}x = t$

so that $\frac{1}{2}\sec^2\frac{1}{2}x\,dx = dt$

$$= 2\cdot\frac{1}{\sqrt{3}}\tan^{-1}\frac{t}{\sqrt{3}} = \frac{2}{\sqrt{3}}\tan^{-1}\left(\frac{1}{\sqrt{3}}\tan\frac{1}{2}x\right).$$

Example 43:

Prove that $\int_0^{\pi} \frac{d\theta}{5+3\cos\theta} = \frac{\pi}{4}$.

Solution:

We have $I = \int_0^{\pi} \frac{d\theta}{5+3\cos\theta}$

$$= \int_0^{\pi} \frac{d\theta}{5\left(\cos^2\frac{1}{2}\theta + \sin^2\frac{1}{2}\theta\right) + 3\left(\cos^2\frac{1}{2}\theta - \sin^2\frac{1}{2}\theta\right)}$$

$$= \int_0^{\pi} \frac{d\theta}{8\cos^2\frac{1}{2}\theta + 2\sin^2\frac{1}{2}\theta} = \frac{1}{2}\int_0^{\pi} \frac{\sec^2\frac{1}{2}\theta\, d\theta}{4+\tan^2\frac{1}{2}\theta}.$$

Now put $\tan\frac{1}{2}\theta = t$

so that $\frac{1}{2}\sec^2\frac{1}{2}\theta\, d\theta = dt$.

Also when $\theta = 0$,

$t = \tan\theta = 0$,

and when $\theta = \pi$,

$t = \tan\frac{1}{2}\pi = \infty$.

$$\therefore\ I = \int_0^{\infty} \frac{dt}{4+t^2} = \frac{1}{2}\left[\tan^{-1}\frac{1}{2}t\right]_0^{\infty}$$

$$= \frac{1}{2}\left[\tan^{-1}\infty - \tan^{-1}0\right] = \frac{1}{2}\left[\frac{\pi}{2} - 0\right] = \frac{\pi}{4}.$$

Example 44:

Evaluate $\int_0^{\pi/2} \frac{d\theta}{1+2\cos\theta}$.

Solution:

We have $I = \int_0^{\pi/2} \frac{d\theta}{1+2\cos\theta}$

$$= \int_0^{\pi/2} \frac{d\theta}{\left(\cos^2\frac{1}{2}\theta + \sin^2\frac{1}{2}\theta\right) + 2\left(\cos^2\frac{1}{2}\theta - \sin^2\frac{1}{2}\theta\right)}.$$

$$= \int_0^{\pi/2} \frac{d\theta}{3\cos^2\frac{1}{2}\theta - \sin^2\frac{1}{2}\theta} = \int_0^{\pi/2} \frac{\sec^2\frac{1}{2}\theta\, d\theta}{3 - \tan^2\frac{1}{2}\theta}.$$

Now put $\tan\frac{1}{2}\theta = t$

so that $\frac{1}{2}\sec^2\frac{1}{2}\theta\, d\theta = dt$.

Also t = 0 when θ = 0

and t = 1

when $\theta = \frac{1}{2}\pi$.

$$\therefore\ I = 2\int_0^1 \frac{dt}{3 - t^2}$$

$$= 2\cdot\frac{1}{2\sqrt{3}}\left[\log\left(\frac{\sqrt{3}+t}{\sqrt{3}-t}\right)_0^1\right] = \frac{1}{\sqrt{3}}\left[\log\left(\frac{\sqrt{3}+1}{\sqrt{3}-1}\right) - \log 1\right]$$

$$= \frac{1}{\sqrt{3}}\log\left(\frac{\sqrt{3}+1}{\sqrt{3}-1}\right) = \frac{1}{\sqrt{3}}\log\left[\frac{(\sqrt{3}+1)(\sqrt{3}+1)}{(\sqrt{3})^2 - 1^2}\right]$$

$$= \frac{1}{\sqrt{3}}\log\left[\frac{4+2\sqrt{3}}{2}\right] = \frac{1}{\sqrt{3}}\log(2+\sqrt{3}).$$

Example 45:

Evaluate $\int_0^{\pi/2} \frac{dx}{1 + a\cos x},\ 0 < a < 1.$

Solution:

Do your self

Example 46:

Prove that $\int_0^{\alpha} \frac{d\theta}{\cos a + \cos\theta} = \operatorname{cosec}\alpha \log(\sec\alpha).$

Solution:

Let $I = \int \frac{d\theta}{\cos\theta + \cos\theta}$

$$= \int \frac{d\theta}{\cos\alpha\left(\cos^2\frac{1}{2}\theta + \sin^2\frac{1}{2}\theta\right) + \left(\cos^2\frac{1}{2}\theta - \sin^2\frac{1}{2}\theta\right)}$$

$$= \int \frac{d\theta}{(1+\cos\alpha)\cos^2\frac{1}{2}\theta - (1-\cos\alpha)\sin^2\frac{1}{2}\theta}$$

$$= \frac{1}{1-\cos\alpha}\int \frac{\sec^2\frac{1}{2}\theta\, d\theta}{\left\{\frac{(1+\cos\alpha)}{(1-\cos\alpha)}\right\} - \tan^2\frac{1}{2}\theta}.$$ **(Note)**

Now $\frac{1+\cos\alpha}{1-\cos\alpha} = \frac{2\cos^2\frac{1}{2}\alpha}{2\sin^2\frac{1}{2}\alpha} = \cot^2\frac{1}{2}\alpha$.

Therefore $I = \frac{1}{2\sin^2\frac{1}{2}\alpha}\int \frac{\sec^2\frac{1}{2}\theta\, d\theta}{\cot^2\frac{1}{2}\alpha - \tan^2\frac{1}{2}\theta}$.

Now putting $\tan\frac{1}{2}\theta = t$

so that $\frac{1}{2}\sec^2\frac{1}{2}\theta\, d\theta = dt$, we get

$$I = \frac{1}{2\sin^2\frac{1}{2}\alpha}\int \frac{2\,dt}{\cot^2\frac{1}{2}\alpha - t^2} = \frac{1}{\sin^2\frac{1}{2}\alpha}\int \frac{dt}{\cot^2\frac{1}{2}\alpha - t^2}$$

$$= \frac{1}{\sin^2\frac{1}{2}\alpha}\cdot\frac{1}{2\cot\frac{1}{2}\alpha}\log\frac{\cot\frac{1}{2}\alpha + t}{\cot\frac{1}{2}\alpha - t}$$

$$= \frac{1}{2\sin\frac{1}{2}\alpha\cos\frac{1}{2}\alpha}\log\left(\frac{\cot\frac{1}{2}\alpha + \tan\frac{1}{2}\theta}{\cot\frac{1}{2}\alpha - \tan\frac{1}{2}\theta}\right)$$

$$= \operatorname{cosec}\alpha \log\left(\frac{\cos\frac{1}{2}\alpha\cos\frac{1}{2}\theta + \sin\frac{1}{2}\alpha\sin\frac{1}{2}\theta}{\cos\frac{1}{2}\alpha\cos\frac{1}{2}\theta - \sin\frac{1}{2}\alpha\sin\frac{1}{2}\theta}\right)$$

$$= \operatorname{cosec}\alpha.\log\left\{\frac{\cos\frac{1}{2}(\alpha-\theta)}{\cos\frac{1}{2}(\alpha+\theta)}\right\}.$$

Hence the given definite integral $\int_0^{\alpha} \frac{d\theta}{\cos\alpha + \cos\theta}$

$$= \operatorname{cosec}\alpha \left[\log \frac{\cos\frac{1}{2}(\alpha-\theta)}{\cos\frac{1}{2}(\alpha+\theta)} \right]_0^{\alpha}$$

$$= \operatorname{cosec}\alpha \left[\log\left(\frac{1}{\cos\alpha}\right) - \log 1 \right]$$

$= \operatorname{cosec}\alpha \log \sec\alpha.$

Example 47:

Prove that $\int_0^{\alpha} \frac{dx}{1-\cos\alpha\cos x} = \frac{\pi}{2}\operatorname{cosec}\alpha.$

Solution:

We have $I = \int_0^{\alpha} \frac{dx}{1-\cos\alpha\cos x}$

$$= \int_0^{\alpha} \frac{dx}{\left(\cos^2\frac{1}{2}x + \sin^2\frac{1}{2}x\right) - \cos\alpha\left(\cos^2\frac{1}{2}x - \sin^2\frac{1}{2}x\right)}$$

$$= \int_0^{\alpha} \frac{dx}{(1-\cos\alpha)\cos^2\frac{1}{2}x + (1+\cos\alpha)\sin^2\frac{1}{2}x}$$

$$= \frac{1}{(1+\cos\alpha)} \int_0^{\alpha} \frac{\sec^2\frac{1}{2}x\,dx}{\left\{\frac{(1-\cos\alpha)}{(1+\cos\alpha)}\right\} + \tan^2\frac{1}{2}x}$$

$$= \frac{1}{2\cos^2\frac{1}{2}\alpha} \int_0^{\alpha} \frac{\sec^2\frac{1}{2}x\,dx}{\tan^2\frac{1}{2}\alpha + \tan^2\frac{1}{2}x}.$$

Now put $\tan\frac{1}{2}x = t$

so that $\frac{1}{2}\sec^2\frac{1}{2}x\,dx = dt$.

When $x = 0$, $t = \tan 0 = 0$

and when $x = \alpha$, $t = \tan\frac{1}{2}\alpha$.

$$\therefore \quad I = \frac{1}{2\cos^2 \frac{1}{2}\alpha} \int_0^{\tan\frac{1}{2}\alpha} \frac{2\,dt}{\tan^2 \frac{1}{2}\alpha + t^2}$$

$$= \frac{1}{\cos^2 \frac{1}{2}\alpha} \int_0^{\tan\frac{1}{2}\alpha} \frac{dt}{\tan^2 \frac{1}{2}\alpha + t^2}$$

$$= \frac{1}{\cos^2 \frac{1}{2}\alpha} \frac{1}{\tan \frac{1}{2}\alpha} \left[\tan^{-1} \frac{t}{\tan\frac{1}{2}\alpha} \right]_0^{\tan\frac{1}{2}\alpha}$$

$$= \frac{1}{\cos\frac{1}{2}\alpha \, \sin\frac{1}{2}\alpha} \left[\tan^{-1} 1 - \tan^{-1} 0\right]$$

$$= \frac{1}{\cos\frac{1}{2}\alpha \, \sin\frac{1}{2}\alpha} \left[\frac{1}{4}\pi - 0\right] = \frac{\pi}{2.\left(2\cos\frac{1}{2}\alpha \, \sin\frac{1}{2}\alpha\right)}$$

$$= \frac{\pi}{2\sin\alpha} = \frac{\pi}{2}\operatorname{cosec}\alpha.$$

Example 48:

Evaluate $\int \frac{\cos\alpha \cos x + 1}{\cos\alpha + \cos x} dx.$

Solution:

$$\text{Given integral } = \int \frac{\cos\alpha \cos x + \left(\cos^2\alpha + \sin^2\alpha\right)}{\cos\alpha + \cos x} dx \qquad \textbf{(Note)}$$

$$= \int \frac{(\cos\alpha + \cos x)\cos\alpha}{(\cos\alpha + \cos x)} dx + \sin^2\alpha \int \frac{dx}{\cos\alpha + \cos x}$$

$$= x\cos\alpha + \sin^2\alpha \operatorname{cosec}\alpha \log \frac{\cos\frac{1}{2}(\alpha - x)}{\cos\frac{1}{2}(\alpha + x)}, \text{ from Ex. 32 (a)}$$

$$= x\cos\alpha + \sin\alpha \log \frac{\cos\frac{1}{2}(\alpha - x)}{\cos\frac{1}{2}(\alpha + x)}.$$

Example 49:

Prove that $\int_0^{\pi} \frac{dx}{1-2a\cos x+a^2} = \frac{\pi}{1-a^2}$ *or* $\frac{\pi}{a^2-1}$ *according as a < or > 1.*

Solution:

We have $I = \int_0^{\pi} \frac{dx}{1-2a\cos x+a^2}$

$$= \int_0^{\pi} \frac{dx}{(1+a^2)\left(\cos^2\frac{1}{2}x+\sin^2\frac{1}{2}x\right)-2a\left(\cos^2\frac{1}{2}x-\sin^2\frac{1}{2}x\right)}$$

$$= \int_0^{\pi} \frac{dx}{(1-a)^2\cos^2\frac{1}{2}x+(1+a)^2\sin^2\frac{1}{2}x}$$

$$= \frac{1}{(1+a)^2}\int_0^{\pi} \frac{\sec^2\left(\frac{1}{2}x\right)dx}{\left\{\frac{(1-a)}{(1+a)}\right\}^2+\tan^2\frac{1}{2}x}$$

$$= \frac{2}{(1+a)^2}\int_0^{\infty} \frac{dt}{\left\{\frac{(1-a)}{(1+a)}\right\}^2+t^2},$$

putting $\tan\frac{1}{2}x = t$

so that $\frac{1}{2}\sec^2\frac{1}{2}x\,dx = dt$

$$= \frac{2}{(1+a)^2}\left[\frac{(1+a)}{(1-a)}\tan^{-1}\left(t\cdot\frac{1+a}{1-a}\right)\right]_0^{\infty}$$

$$= \frac{2}{(1+a)(1-a)}\left[\tan^{-1}\left(t\cdot\frac{1+a}{1-a}\right)\right]_0^{\infty}$$

$= \frac{2}{1-a^2}\left[\tan^{-1}\infty-\tan^{-1}0\right]$, if a < 1 [$\because$ a < 1 means 1 – a is + ive]

$$= \frac{2}{1-a^2}\left[\frac{1}{2}\pi-0\right] = \frac{\pi}{1-a^2}.$$

If a > 1, then $I = \frac{2}{1-a^2}\left[\tan^{-1}\left(t \cdot \frac{1+a}{1-a}\right)\right]_0^{\infty}$

$= \frac{2}{1-a^2}\left[\tan^{-1}(-\infty) - \tan^{-1} 0\right]$ [∵ a > 1 means 1 – a is – ive]

$= \frac{2}{1-a^2}\left[-\frac{1}{2}\pi - 0\right] = \frac{\pi}{a^2-1}$.

Example 50:

Evaluate $\int_0^{\pi/2} \frac{dx}{a^2\cos^2 x + b^2 \sin^2 x}$.

Solution:

We have

$$I = \int_0^{\pi/2} \frac{dx}{a^2\cos^2 x + b^2\sin^2 x} = \int_0^{\pi/2} \frac{\sec^2 x\, dx}{a^2 + b^2 \tan^2 x},$$

dividing the numerator and the denominator by $\cos^2 x$.

Now put b tan x = t

so that b $\sec^2$ x dx = dt.

When $x = \frac{1}{2}\pi$, $t = b\tan\frac{1}{2}\pi = \infty$

and when x = 0, t = b tan 0 = 0.

$$\therefore\ I = \frac{1}{b}\int_0^{\infty} \frac{dt}{a^2+t^2} = \frac{1}{b}\cdot\frac{1}{a}\left[\tan^{-1}\frac{t}{a}\right]_0^{\infty}$$

$$= \frac{1}{ab}\left[\tan^{-1}\infty - \tan^{-1} 0\right] = \frac{1}{ab}\left[\frac{\pi}{2} - 0\right] = \frac{\pi}{2ab}.$$

Example 51:

Evaluate $\int \frac{\sin 2x\, dx}{(a + b\cos x)^2}$.

Solution:

We have $I = \int \frac{\sin 2x\, dx}{(a+b\cos x)^2} = 2\int \frac{\sin x \cos x\, dx}{(a+b\cos x)^2}$.

Now put a + b cos x = t

so that – b sin x dx = dt.

Also cos x = $\frac{(t-a)}{b}$.

$$\therefore\ I = -\frac{2}{b}\int \frac{(t-a)/b}{t^2}dt = -\frac{2}{b^2}\int\left[\frac{t}{t^2}-\frac{a}{t^2}\right]dt$$

$$= -\frac{2}{b^2}\int\left[\frac{1}{t}-\frac{a}{t^2}\right]dt = -\frac{2}{b^2}\left[\log t+\frac{a}{t}\right]$$

$$= -\frac{2}{b^2}\left[\log(a+b\cos x)+\frac{a}{a+b\cos x}\right].$$

Example 52:

Evaluate $\int \frac{\cos x}{a+b\cos x}dx.$

Solution:

We have $\int \frac{\cos x\,dx}{a+b\cos x} = \frac{1}{b}\int \frac{b\cos x\,dx}{a+b\cos x}$ **(Note)**

$$= \frac{1}{b}\int \frac{a+b\cos x-a}{(a+b\cos x)}dx = \frac{1}{b}\int\left(1-\frac{a}{a+b\cos x}\right)dx$$

$$= \frac{1}{b}x-\frac{a}{b}\int \frac{1}{a+b\cos x}dx.$$

Example 53:

Evaluate $\int \frac{dx}{1+3\sin^2 x}.$

Solution:

We have $\int \frac{dx}{1+3\sin^2 x} = \int \frac{dx}{(\sin^2 x+\cos^2 x)+3\sin^2 x}$ **(Note)**

$$= \int \frac{dx}{4\sin^2 x+\cos^2 x} = \int \frac{\sec^2 x\,dx}{4\tan^2 x+1},$$

dividing the numerator and the denominator by $\cos^2 x$

$$= \frac{1}{2}\int \frac{dt}{t^2+1},\ \text{putting } 2\tan x = t$$

so that $2\sec^2 x\,dx = dt$

$$= \frac{1}{2}\cdot\tan^{-1}(t) = \frac{1}{2}\tan^{-1}(2\tan x).$$

Example 54:

Evaluate $\int \frac{dx}{a^2-b^2\cos^2 x},\ a > b.$

Solution:

We have $\int \frac{dx}{a^2 - b^2 \cos^2 x}$

$$= \int \frac{dx}{a^2(\sin^2 x + \cos^2 x) - b^2 \cos^2 x} \qquad \textbf{(Note)}$$

$$= \int \frac{dx}{(a^2 - b^2)\cos^2 x + a^2 \sin^2 x} = \frac{1}{a^2}\int \frac{\sec^2 x\,dx}{\left\{\frac{(a^2 - b^2)}{a^2}\right\} + \tan^2 x}$$

$$= \frac{1}{a^2}\int \frac{dt}{\left\{\frac{(a^2 - b^2)}{a^2}\right\} + t^2}$$, putting tan x = t so that $\sec^2 x\ dx = dt$

$$= \frac{1}{a^2} \cdot \frac{a}{\sqrt{(a^2 - b^2)}} \tan^{-1}\left\{t \cdot \frac{a}{\sqrt{(a^2 - b^2)}}\right\}$$

$$= \frac{1}{a\sqrt{(a^2 - b^2)}} \tan^{-1}\left\{\tan x \cdot \frac{a}{\sqrt{(a^2 - b^2)}}\right\}.$$

Example 55:

Evaluate $\int \frac{dx}{1 + \cos^2 x}$.

Solution:

We have $\int \frac{dx}{1 + \cos^2 x} = \int \frac{\sec^2 x\,dx}{\sec^2 x + 1}$,

dividing the numerator and the denominator by $\cos^2 x$

$$= \int \frac{\sec^2 x\,dx}{1 + \tan^2 x + 1} = \int \frac{\sec^2 x\,dx}{2 + \tan^2 x}$$

$$= \int \frac{dt}{2 + t^2},$$

putting tan x = t

so that $\sec^2 x\ dx = dt$

$$= \frac{1}{\sqrt{2}} \tan^{-1}\left(\frac{t}{\sqrt{2}}\right) = \frac{1}{\sqrt{2}} \tan^{-1}\left(\frac{1}{\sqrt{2}} \tan x\right).$$

Example 56:

Evaluate $\int \frac{dx}{a+b\cosh x}$.

Solution:

Use $\cosh x = \cosh^2 \frac{1}{2}x + \sinh^2 \frac{1}{2}x$ and

$\cosh^2 \frac{1}{2}x - \sinh^2 \frac{1}{2}x = 1$. Discuss both the cases *i.e.*, when $a < b$ and when $a > b$.

Here we get $\int \frac{dx}{a+b\cosh x}$

$$= -\frac{1}{\sqrt{(a^2-b^2)}} \log\left\{\frac{\sqrt{(a-b)}\tanh\frac{1}{2}x - \sqrt{(a+b)}}{\sqrt{(a-b)}\tanh\frac{1}{2}x + \sqrt{(a+b)}}\right\}, \text{ when } a > b$$

or $$= \frac{2}{\sqrt{(b^2-a^2)}} \tan^{-1}\left\{\sqrt{\left(\frac{b-a}{b+a}\right)}\tanh\frac{1}{2}x\right\}, \text{ when } a < b.$$

Example 57:

Evaluate $\int_0^{\pi/2} \frac{dx}{5+4\sin x}$.

Solution:

We have $\int_0^{\pi/2} \frac{dx}{5+4\sin x}$

$$= \int_0^{\pi/2} \frac{dx}{5\left(\cos^2\frac{1}{2}x + \sin^2\frac{1}{2}x\right) + 4.2\sin\frac{1}{2}x\cos\frac{1}{2}x}$$

$$= \int_0^{\pi/2} \frac{\sec^2\frac{1}{2}x\,dx}{5+8\tan\frac{1}{2}x + 5\tan^2\frac{1}{2}x}, \quad \text{dividing Nr. and Dr. by } \cos^2\frac{1}{2}x$$

$$= \int_0^1 \frac{2\,dt}{5+8t+5t^2}, \quad \left[\text{putting } \tan\frac{1}{2}x = t \text{ and changing the limits}\right]$$

$$= \int_0^1 \frac{2\,dt}{5\left(t^2+\frac{8}{5}t+1\right)} = \int_0^1 \frac{2\,dt}{5\left[\left(t+\frac{4}{5}\right)^2+\frac{9}{25}\right]}$$

$$= \frac{2}{5}\left[\frac{5}{3}\tan^{-1}\left\{\frac{\left(t+\frac{4}{5}\right)}{\frac{3}{5}}\right\}\right]_0^1 = \frac{2}{3}\left[\tan^{-1}\frac{5t+4}{3}\right]_0^1$$

$$= \frac{2}{3}\left[\tan^{-1}3 - \tan^{-1}\frac{4}{3}\right].$$

Example 58:

Evaluate $\int_0^{\pi/2} \frac{dx}{4+5\sin x}$.

Solution:

We have $I = \int_0^{\pi/2} \frac{dx}{4+5\sin x}$

$$= \int_0^{\pi/2} \frac{dx}{4\left(\cos^2\frac{1}{2}x + \sin^2\frac{1}{2}x\right) + 5.2\sin\frac{1}{2}x\cos\frac{1}{2}x}$$

$$= \frac{1}{2}\int_0^{\pi/2} \frac{dx}{2\cos^2\frac{1}{2}x + 5\sin\frac{1}{2}\cos\frac{1}{2}x + 2\sin^2\frac{1}{2}x}$$

$$= \frac{1}{2}\int_0^{\pi/2} \frac{\sec^2\frac{1}{2}x\,dx}{2+5\tan\frac{1}{2}x + 2\tan^2\frac{1}{2}x}$$ dividing Nr. and Dr. by $\cos^2\frac{1}{2}x$.

Now put $\tan\frac{1}{2}x = t$

so that $\frac{1}{2}\sec^2\frac{1}{2}x\,dx = dt$. When $x = 0$,
$t = \tan 0 = 0$,

when $x = \frac{\pi}{2}$,

$t = \tan\frac{1}{4}\pi = 1$.

$$\therefore\ I = \int_0^1 \frac{dt}{2t^2+5t+2}$$

$$= \frac{1}{2}\int_0^1 \frac{dt}{\left(t^2+\frac{5}{2}t+1\right)} = \frac{1}{2}\int_0^1 \frac{dt}{\left(t+\frac{5}{4}\right)^2 - \frac{9}{16}}$$

$$=\frac{1}{2}\frac{1}{2.\left(\frac{3}{4}\right)}\left\{\log\frac{\left(t+\frac{5}{4}\right)-\frac{3}{4}}{\left(t+\frac{5}{4}\right)+\frac{3}{4}}\right\}_0^1=\frac{1}{3}\left[\log\frac{t+\frac{1}{2}}{t+\frac{8}{4}}\right]_0^1$$

$$=\frac{1}{3}\left[\log\frac{2t+1}{2t+4}\right]_0^1$$

$$=\frac{1}{3}\left(\log\frac{1}{2}-\log\frac{1}{4}\right)=\frac{1}{3}\log\left(\frac{1}{2}\div\frac{1}{4}\right)$$

$$=\frac{1}{3}\log\left(\frac{1}{2}\times 4\right)=\frac{1}{3}\log 2.$$

Example 59:

Evaluate $\int\frac{dx}{3\sin x+4\cos x}$.

Solution:

Given integral $=\int\frac{dx}{3\cdot 2\sin\frac{1}{2}x\cos\frac{1}{2}x+4\left(\cos^2\frac{1}{2}x-\sin^2\frac{1}{2}x\right)}$

$$=\int\frac{\sec^2\frac{1}{2}x\,dx}{4+6\tan\frac{1}{2}x-4\tan^2\frac{1}{2}x}$$

dividing the Nr. and the Dr. by $\cos^2\frac{1}{2}x$

$$=\int\frac{dt}{2+3t-2t^2},$$

putting $\tan\frac{1}{2}x=t$

so that $\frac{1}{2}\sec^2\frac{1}{2}x\,dx=dt$

$$=\frac{1}{2}\int\frac{dt}{1-\left(t^2-\frac{3}{2}t\right)}=\frac{1}{2}\int\frac{dt}{1-\left(t-\frac{3}{4}\right)^2+\frac{9}{16}}$$

$$=\frac{1}{2}\int\frac{dt}{\frac{25}{16}-\left(t-\frac{3}{4}\right)^2}=\frac{1}{2}\cdot\frac{1}{2}\cdot\frac{4}{5}\log\frac{\frac{5}{4}+\left(t-\frac{3}{4}\right)}{\frac{5}{4}-\left(t-\frac{3}{4}\right)}$$

$$=\frac{1}{5}\log\frac{1+2t}{4-2t}=\frac{1}{5}\log\frac{1+2\tan\frac{1}{2}x}{4-2\tan\frac{1}{2}x}.$$

Example 60:

Evaluate $\int\frac{dx}{5\sin x+12\cos x}$.

Solution:

The given integral

$$=\int\frac{dx}{5.2\sin\frac{1}{2}x\cos\frac{1}{2}x+12\left(\cos^2\frac{1}{2}x-\sin^2\frac{1}{2}x\right)}$$

$$=\frac{1}{12}\int\frac{\sec^2\frac{1}{2}x\,dx}{1+\frac{5}{6}\tan\frac{1}{2}x-\tan^2\frac{1}{2}x}=\frac{1}{6}\int\frac{dt}{1+\frac{5}{6}t-t^2}$$

putting $\tan\frac{1}{2}x=t$ so that $\frac{1}{2}\sec^2\frac{1}{2}x\,dx=dt$

$$=\frac{1}{6}\int\frac{dt}{\frac{169}{144}-\left(t-\frac{5}{12}\right)^2}=\frac{1}{6}\cdot\frac{1}{2}\cdot\frac{12}{13}\log\frac{\frac{13}{12}+\left(t-\frac{5}{12}\right)}{\frac{13}{12}-\left(t-\frac{5}{12}\right)}$$

$$=\frac{1}{13}\log\left\{\frac{\frac{8}{12}+t}{\frac{18}{12}-t}\right\}=\frac{1}{13}\log\left(\frac{8+12t}{18-12t}\right)$$

$$=\frac{1}{13}\log\left\{\frac{2}{3}\left(\frac{2+3t}{3-2t}\right)\right\}=\frac{1}{13}\log\left\{\frac{2}{3}\left(\frac{2+3\tan\frac{1}{2}x}{3-2\tan\frac{1}{2}x}\right)\right\}.$$

Example 61:

Show that $\int_0^{\pi/2}\frac{dx}{1+2\sin x+\cos x}=\frac{1}{2}\log 3$.

Solution:

The given integral $I=\int_0^{\pi/2}\frac{dx}{(1+\cos x)+2\sin x}$

$$=\int_0^{\pi/2}\frac{dx}{2\cos^2\frac{1}{2}x+4\sin\frac{1}{2}x\cos\frac{1}{2}x}$$

$$= \int_0^{\pi/2} \frac{\sec^2 \frac{1}{2} x \, dx}{2\left(1 + 2\tan\frac{1}{2}x\right)}, \text{ dividing the Nr. and the Dr. by } \cos^2 \frac{1}{2}x$$

$$= \int_0^1 \frac{dt}{1+2t}, \text{ putting } \tan\frac{x}{2} - t$$

so that $\frac{1}{2}\sec^2\frac{x}{2}dx = dt$

$$= \frac{1}{2}[\log(1+2t)]_0^1 = \frac{1}{2}(\log 3 - \log 1) = \frac{1}{2}\log 3.$$

Example 62:

Evaluate $\int \frac{3\sin x + 4\cos x}{\sin x + \cos x} dx.$

Solution:

Here we put

Numerator ≡ *A. (deno.) + B. (diff. coeff. of deno.)*

i.e., 3 sin x + 4 cos x ≡ A (sin x + cos x) + B (cos x − sin x).

Equating the coefficients of sin x and cos x on both sides, we have

$$3 = A - B \text{ and } 4 = A + B;$$

whence $A = \frac{7}{2}$

and $B = \frac{1}{2}$.

∴ the given integral

$$= \int \frac{\frac{7}{2}(\sin x + \cos x) + \frac{1}{2}(\cos x - \sin x)}{\sin x + \cos x} dx$$

$$= \frac{7}{2}\int dx + \frac{1}{2}\int \frac{\cos x - \sin x}{\sin x + \cos x} dx = \frac{7}{2}x + \frac{1}{2}\log(\sin x + \cos x).$$

Example 63:

Evaluate $\int \frac{\sin x + 2\cos x}{2\sin x + \cos x} dx.$

Solution:

Here we put

(sin x + 2 cos x) ≡ A (2 sin x + cos x) + B (2 cos x − sin x).

Equating the coefficients of sin x and cos x on both the sides, we have 1 = 2A − B

and 2 = A + 2B;

whence $A = \frac{4}{5}$

and $B = \frac{3}{5}$.

∴ the given integral

$$= \int \frac{\frac{4}{5}(2\sin x + \cos x) + \frac{3}{5}(2\cos x - \sin x)}{2\sin x + \cos x} dx$$

$$= \frac{4}{5}\int dx + \frac{3}{5}\int \frac{(2\cos x - \sin x)}{2\sin x + \cos x} dx = \frac{4}{5}x + \frac{3}{5}\log(2\sin x + \cos x).$$

Example 64:

Evaluate $\int \frac{(2\cos x - \sin x)}{2\sin x + \cos x} dx$.

Solution:

Here we put

2 sin x + 3 cos x ≡ A (3 sin x + 4 cos x) + B (3 cos x − 4 sin x).

Equating the coefficients of sin x and cos x on both the sides, we have

2 = 3A − 4B

and 3 = 4A + 3B;

whence $A = \frac{18}{25}$

and $B = \frac{1}{25}$.

∴ the given integral

$$= \int \frac{\frac{18}{25}(3\sin x + 4\cos x) + \frac{1}{25}(3\cos x - 4\sin x)}{3\sin x + 4\cos x} dx$$

$$= \frac{18}{25}\int dx + \frac{1}{25}\int \frac{3\cos x - 4\sin x}{3\sin x + 4\cos x} dx$$

$$= \frac{18}{25}x + \frac{1}{25}\log(3\sin x + 4\cos x).$$

Example 65:

Show that $\int_0^{\pi} \frac{dx}{3+2\sin x+\cos x} = \frac{\pi}{4}$.

Solution:

The given integral

$$= \int_0^{\pi} \frac{dx}{3\left(\cos^2 \frac{1}{2}x + \sin^2 \frac{1}{2}x\right) + 2\left(2\sin\frac{1}{2}x \cos\frac{1}{2}x\right) + \left(\cos^2 \frac{1}{2}x - \sin^2 \frac{1}{2}x\right)}$$

$$= \int_0^{\pi} \frac{dx}{4\cos^2 \frac{1}{2}x + 2\sin^2 \frac{1}{2}x + 4\sin\frac{1}{2}x \cos\frac{1}{2}x}$$

$$= \int_0^{\pi} \frac{\sec^2\left(\frac{1}{2}x\right)dx}{4 + 2\tan^2 \frac{1}{2}x + 4\tan\frac{1}{2}x},$$

dividing the Nr. and the Dr. by $\cos^2 \frac{1}{2}x$

$$= \int_0^{\pi} \frac{\frac{1}{2}\sec^2 \frac{1}{2}x\,dx}{\tan^2 \frac{1}{2}x + 2\tan\frac{1}{2}x + 2}$$

$$= \int_0^{\infty} \frac{dt}{t^2 + 2t + 2},$$ putting $\tan\frac{1}{2}x = t$ and changing the limits

$$= \int_0^{\infty} \frac{dt}{(t+1)^2 + 1} = \left[\tan^{-1}(t+1)\right]_0^{\infty}$$

$$= \tan^{-1}\infty - \tan^{-1}1 + \frac{\pi}{2} - \frac{\pi}{4} = \frac{\pi}{4}.$$

Example 66:

Evaluate $\int \frac{17\cos x - 6\sin x}{3\sin x + 4\cos x}$.

Solution:

Do your self.

The given integral = 2x + 3 log (3 sin x + 4 cos x).

Example 67:

Evaluate $\int \frac{\cos x\,dx}{2\sin x + 3\cos x}$.

Solution:

Here we put

Numerator ≡ A.(deno.) + B. (diff. coeff. of deno.)

i.e., cos ≡ A (2 sin x + 3 cos x) + B (2 cos x − 3 sin x).

Equating the coefficients of sin x and cos x on both the sides, we have

0 = 2A − 3B and 1 = 3A + 2B;

whence $A = \frac{3}{13}$ and $B = \frac{2}{13}$.

∴ the given integral

$$= \int \frac{\frac{3}{12}(2\sin x + 3\cos x) + \frac{2}{13}(2\cos x - 3\sin x)}{2\sin x + 3\cos x} dx$$

$$= \frac{3}{13}\int dx + \frac{2}{13}\int \frac{2\cos x - 3\sin x}{2\sin x + 3\cos x} dx$$

$$= \frac{3}{13}x + \frac{2}{13}\log(2\sin x + 3\cos x).$$

Example 68:

Evaluate $\int \frac{\cos x\, dx}{3\cos x + 4\sin x}$.

Solution:

Do your self.

The given integral $= \frac{3}{25}x + \frac{4}{25}\log(3\cos x + 4\sin x)$.

Example 69:

Evaluate $\int \frac{3 + 4\sin x + 2\cos x}{3 + 2\sin x + \cos x} dx$.

Solution:

Here we put

Numerator ≡ A. (deno.) + B. (diff. coeff. of deno.) + C.

Thus let 3 + 4 sin x + 2 cos x

≡ A (3 + 2 sin x + cos x) + B. (2 cos x − sin x) + C. ...(1)

Equating the coefficients of sin x, cos x and constant terms on both sides, we have 4 = 2A − B, = A + 2B, and 3 = C + 3A.

Solving these equations, we have

A = 2, B = 0 and C = – 3.

∴ the given integral

$$= \int \frac{2(3+2\sin x+\cos x)dx}{3+2\sin x+\cos x} - \int \frac{3dx}{3+2\sin x+\cos x}$$

$$= \int 2dx$$

$$- \int \frac{3dx}{3\left(\cos^2\frac{1}{2}x+\sin^2\frac{1}{2}x\right)+4\sin\frac{1}{2}x\cos\frac{1}{2}x+\cos^2\frac{1}{2}x-\sin^2\frac{1}{2}x}$$

$$= 2\int dx - \int \frac{3dx}{4\cos^2\frac{1}{2}x+4\sin\frac{1}{2}x\cos\frac{1}{2}x+2\sin^2\frac{1}{2}x}$$

$$= 2x - \int \frac{3\sec^2\frac{1}{2}x\,dx}{4+4\tan\frac{1}{2}x+2\tan^2\frac{1}{2}x}$$

$$= 2x - 3\int \frac{\frac{1}{2}\sec^2\frac{1}{2}x\,dx}{2+2\tan\frac{1}{2}x+\tan^2\frac{1}{2}x}$$

$$= 2x - 3\int \frac{dt}{t^2+2t+2},$$

putting $\tan\frac{1}{2}x = t$

so that $\frac{1}{2}\sec^2\frac{1}{2}x\,dx = dt$

$$= 2x - 3\int \frac{dt}{(t+1)^2+1} = 2x - 3\tan^{-1}(t+1)$$

$$= 2x - 3\tan^{-1}\left(1+\tan\frac{1}{2}x\right).$$

Example 70:

Evaluate $\int \frac{\sin x+\cos x}{3\sin x+4\cos x+1}dx$.

Solution:

Let sin x + cos x ≡ A (3 sin x + 4 cos x + 1)

+ B (3 cos x – 4 sin x) + C.

Equating the coefficients of sin x, cos x and constant terms on both sides, we have

$$1 = 3A - 4B,$$

$$1 = 4A + 3B,$$

and $\quad 0 = A + C,$

Solving, we have $A = \frac{7}{25}$,

$$B = -\frac{1}{25}$$

and $\quad C = -A = -\frac{7}{25}.$

$\therefore$ the given integral $= \frac{7}{25}\int dx - \frac{1}{25}\int \frac{3\cos x - 4\sin x}{3\sin x + 4\cos x + 1}dx$

$$-\frac{7}{25}\int \frac{dx}{3\sin x + 4\cos x + 1}$$

$$= \frac{7}{25}x - \frac{1}{25}\log(3\sin x + 4\cos x + 1) - \frac{7}{25}I, \qquad \textbf{(Note)}$$

where $I = \int \frac{dx}{3\sin x + 4\cos x + 1}$

$$= \int \frac{dx}{5\cos^2\frac{1}{2}x + 6\sin\frac{1}{2}x\cos\frac{1}{2}x - 3\sin^2\frac{1}{2}x} \qquad \textbf{(Note)}$$

$$= \int \frac{\sec^2\frac{1}{2}x\,dx}{5 + 6\tan\frac{1}{2}x - 3\tan^2\frac{1}{2}x} = \int \frac{2\,dt}{5 + 6t - 3t^2},$$

putting $\tan\frac{1}{2}x = t$

so that $\frac{1}{2}\sec^2\frac{1}{2}x\,dx = dt$

$$= \frac{1}{3}\int \frac{dt}{\frac{8}{3} - (t-1)^2} = \frac{1}{3}\cdot\frac{1}{2.\sqrt{\left(\frac{8}{3}\right)}}\log\frac{\sqrt{\left(\frac{8}{3}\right)} + (t-1)}{\sqrt{\left(\frac{8}{3}\right)} - (t-1)}$$

$$= \frac{1}{4\sqrt{6}}\log\frac{\sqrt{\left(\frac{8}{3}\right)} + 1 + \tan\frac{1}{2}x}{\sqrt{\left(\frac{8}{3}\right)} + 1 - \tan\frac{1}{2}x}.$$

Example 71:

Evaluate $\int \frac{2+3\cos x}{\sin x+2\cos x+3}dx.$

Solution:

Do yourself.

Ans. $\frac{6}{5}x+\frac{3}{5}\log(\sin x+2\cos x+3)-\frac{8}{5}\tan^{-1}\left\{\frac{1}{3}\left(1+\tan\frac{1}{2}x\right)\right\}$

Example 72:

Evaluate $\int \cos 2x \log(1+\tan x)dx.$

Solution:

Integrating by parts taking cos 2x as the 2nd function, the given integral

$$=\{\log(1+\tan x)\}\frac{\sin 2x}{2}-\int\frac{\sec^2 x}{1+\tan x}\cdot\frac{\sin 2x}{2}dx$$

$$=\frac{1}{2}\sin 2x \log(1+\tan x)-\int\frac{\sin x}{\sin x+\cos x}dx.$$

Now $\int\frac{\sin x\,dx}{\sin x+\cos x}$

$$=\frac{1}{2}\int\frac{(\sin x+\cos x)-(\cos x-\sin x)}{\sin x+\cos x}dx, \qquad \textbf{(Note)}$$

$$=\frac{1}{2}\int\left[1-\frac{\cos x-\sin x}{\sin x+\cos x}\right]dx=\frac{1}{2}[x-\log(\sin x+\cos x)].$$

Hence the given integral

$$=\frac{1}{2}\sin 2x \log(1+\tan x)-\frac{1}{2}[x-\log(\sin x+\cos x)].$$

Example 73:

Show that $\int_0^{\pi}\frac{dx}{3+2\sin x+\cos x}=\frac{\pi}{4}.$

Solution:

The given integral

$$=\int_0^{\pi}\frac{dx}{3\left(\cos^2\frac{1}{2}x+\sin^2\frac{1}{2}x\right)+2\left(2\sin\frac{1}{2}x\cos\frac{1}{2}x\right)+\left(\cos^2\frac{1}{2}x-\sin^2\frac{1}{2}x\right)}$$

$$=\int_0^{\pi}\frac{dx}{4\cos^2\frac{1}{2}x+2\sin^2\frac{1}{2}x+4\sin\frac{1}{2}x\cos\frac{1}{2}x}$$

$$=\int_0^{\pi}\frac{\sec^2\left(\frac{1}{2}x\right)dx}{4+2\tan^2\frac{1}{2}x+4\tan\frac{1}{2}x},$$

dividing the Nr. and the Dr. by $\cos^2\frac{1}{2}x$

$$=\int_0^{\pi}\frac{\frac{1}{2}\sec^2\frac{1}{2}x\,dx}{\tan^2\frac{1}{2}x+2\tan\frac{1}{2}x+2}$$

$$=\int_0^{\infty}\frac{dt}{t^2+2t+2},$$ putting $\tan\frac{1}{2}x=t$ and changing the limits

$$=\int_0^{\infty}\frac{dt}{(t+1)^2+1}=\left[\tan^{-1}(t+1)\right]_0^{\infty}$$

$$=\tan^{-1}\infty-\tan^{-1}1+\frac{\pi}{2}-\frac{\pi}{4}=\frac{\pi}{4}.$$

Example 74:

Evaluate $\int\frac{\sin x}{\sin x+\cos x}dx$.

Solution:

Do your self. **Ans.** $\frac{1}{2}[x-\log(\sin x+\cos x)]$.

Example 75:

Evaluate $\int\frac{1}{1+\tan x}dx$.

Solution:

The given integral $I=\int\frac{1}{1+\left(\frac{\sin x}{\cos x}\right)}dx$

$$=\int\frac{\cos x}{(\cos x+\sin x)}dx$$

$$=\frac{1}{2}\int\frac{(\cos x+\sin x)+(\cos x-\sin x)}{\cos x+\sin x}dx$$

$$=\frac{1}{2}\int\left[1+\frac{\cos x-\sin x}{\sin x+\cos x}\right]dx$$

$$=\frac{1}{2}[x+\log(\sin x+\cos x)].$$

Example 76:

Evaluate $\int\frac{dx}{x+\sqrt{(a^2+x^2)}}$.

Solution:

Let $I=\int\frac{dx}{x+\sqrt{(a^2-x^2)}}$.

Put $x = a\sin\theta$,

so that $dx = a\cos\theta\, d\theta$.

Then $I=\int\frac{a\cos\theta\, d\theta}{a\sin\theta+a\cos\theta}=\int\frac{\cos\theta}{\sin\theta+\cos\theta}d\theta$

$=\frac{1}{2}[\theta+\log(\sin\theta+\cos\theta)]$, proceeding as in above ex.

$$=\frac{1}{2}\theta+\frac{1}{2}\log\left[\sin\theta+\sqrt{(1-\sin^2\theta)}\right]$$

$$=\frac{1}{2}\sin^{-1}\frac{x}{a}+\frac{1}{2}\log\left[\frac{x}{a}+\sqrt{\left(1-\frac{x^2}{a^2}\right)}\right]$$

$$=\frac{1}{2}\sin^{-1}\frac{x}{a}+\frac{1}{2}\log\left[x+\sqrt{(a^2-x^2)}\right]-\frac{1}{2}\log a$$

$=\frac{1}{2}\sin^{-1}\frac{x}{a}+\frac{1}{2}\log\left[x+\sqrt{(a^2-x^2)}\right]$, the constant term $-\frac{1}{2}\log a$ may be added to the constant of integration c.

Example 77:

Integrate $\cos^5 x$.

Solution:

$$\int\cos^5 x\,dx=\int\cos^4 x\cos x\,dx=\int(1-\sin^2 x)^2\cos x\,dx$$

$=\int(1-t^2)^2\,dt$, [putting $\sin x = t$ so that $\cos x\,dx = dt$]

$= \int (1 - 2t^2 + t^4) dt = t - \frac{2}{3}t^3 + \frac{1}{5}t^5.$

$\therefore \int \cos^5 x \, dx = \sin x - \frac{2}{3}\sin^3 x + \frac{1}{5}\sin^5 x.$

Similarly, $\int \sin^5 x \, dx = -\cos x + \frac{2}{3}\cos^3 x - \frac{1}{5}\cos^5 x.$

Example 78:

Integrate $\sin^7 x$.

Solution:

$\int \sin^7 x \, dx = \int \sin^6 x . \sin x \, dx = \int (1 - \cos^2 x)^3 . \sin x \, dx$

$= -\int (1 - t^2)^3 dt,$ [putting cos x = t so that – sin x dx = dt]

$= -\int (1 - 3t^2 + 3t^4 - t^6) dt = -t + t^3 - \frac{3}{5}t^5 + \frac{1}{7}t^7.$

$\therefore \int \sin^7 x \, dx = -\cos x + \cos^3 x - \frac{3}{5}\cos^5 x + \frac{1}{7}\cos^7 x.$

Example 79:

Integrate $\cos^7 x$.

Solution:

$\int \cos^7 x \, dx = \int \cos^6 x . \cos x \, dx = \int (1 - \sin^2 x)^3 \cos x \, dx$

$= \int (1 - t^2)^3 dt,$ [putting sin x = t so that cos x dx = dt]

$= \int (1 - 3t^2 - 3t^4 - t^6) dt = t - t^3 + \frac{3}{5}t^5 - \frac{1}{7}t^7$

$= \sin x - \sin^3 x + \frac{3}{5}\sin^5 x - \frac{1}{7}\sin^7 x.$

Example 80:

Integrate $\cos^4 x$.

Solution:

We have $\cos^4 x = (\cos^2 x)^2 = \left\{\frac{1}{2}(1 + \cos 2x)\right\}^2$

$= \frac{1}{4}(1 + \cos 2x)^2 = \frac{1}{4}(1 + 2\cos 2x + \cos^2 2x)$

$$=\frac{1}{4}\left[1+2\cos 2x+\frac{1}{2}(1+\cos 4x)\right]$$

$$=\frac{1}{8}[3+4\cos 2x+\cos 4x].$$

$$\therefore \int \cos^4 x\,dx=\frac{1}{8}\left[\int 3dx+4\int \cos 2x\,dx+\int \cos 4x\,dx\right]$$

$$=\frac{1}{8}\left[3x+4\cdot\frac{1}{2}\sin 2x+\frac{1}{4}\sin 4x\right]$$

$$=\frac{3}{8}x+\frac{1}{4}\sin 2x+\frac{1}{32}\sin 4x.$$

Example 81:

Evaluate $\int_0^{\pi/4} \sin^4 x\,dx.$

Solution:

We have $\sin^4 x = (\sin^2 x)^2 = \left[\frac{1}{2}(1-\cos 2x)\right]^2$

$$=\frac{1}{4}[1-2\cos 2x+\cos^2 2x]$$

$$=\frac{1}{4}\left[1-2\cos 2x+\frac{1}{2}(1+\cos 4x)\right]$$

$$=\frac{1}{8}[3-4\cos 2x+\cos 4x].$$

$$\therefore \int_0^{\pi/4} \sin^4 x\,dx=\frac{1}{8}\int_0^{\pi/4}[3-4\cos 2x+\cos 4x]dx$$

$$=\frac{1}{8}\left[3x-\frac{4\sin 2x}{2}+\frac{\sin 4x}{4}\right]_0^{\pi/4}$$

$$=\frac{1}{8}\left[\left\{3\cdot\frac{\pi}{4}-2\sin\frac{\pi}{2}+\frac{1}{4}\sin\pi\right\}-0\right]=\frac{1}{32}[3\pi-8].$$

2

REDUCTION FORMULAE

(Trigonometric Functions)

2.0. INTRODUCTION

A reduction formula is a formula which connects an integral, which cannot otherwise be evaluated, with another integral of the same type but of lower degree. It is generally obtained by applying the rule of integration by parts.

2.1. REDUCTION FORMULAE FOR

$\int \sin^n x\,dx$ and $\int \cos^n x\,dx$, n being a +ive integer

(a) Let $I_n = \int \sin^n x\,dx$ or $I_n = \int \sin^{n-1} x \sin x\,dx$. **(Note)**

Integrating by parts regarding sin x as the 2nd function, we have

$$I_n = \sin^{n-1} x.(-\cos x) - \int (n-1) \sin^{n-2} x.\cos x.(-\cos x)dx$$

$$= -\sin^{n-1} x.\cos x + (n-1)\int \sin^{n-2} x.\cos^2 x\,dx$$

$$= -\sin^{n-1} x.\cos x + (n-1)\int \sin^{n-2} x.\left(1-\sin^2 x\right)dx \qquad \textbf{(Note)}$$

$$= -\sin^{n-1} x.\cos x + (n-1)\int \sin^{n-2} x\,dx - (n-1)\int \sin^n x\,dx$$

$$= -\sin^{n-1} x.\cos x + (n-1)\int \sin^{n-2} x\,dx - (n-1)I_n.$$

Transposing the last term to the left, we have

$$I_n (1 + n - 1) = -\sin^{n-1} x.\cos x + (n-1) I_{n-2},$$

$$\left[\because I_{n-2} = \int \sin^{n-2} x\,dx\right]$$

or $nI_n = -\sin^{n-1} x \cos x + (n-1) I_{n-2}$

or $I_n = -\dfrac{\sin^{n-1} x \cos x}{n} + \dfrac{n-1}{n} I_{n-2}$.

$$\therefore \int \sin^n x\,dx = -\frac{1}{n}\sin^{n-1} x.\cos x + \frac{n-1}{n}\int \sin^{n-2} x\,dx.$$

(b) Let $I_n = \int \cos^n x\,dx$ or $I_n = \int \cos^{n-1}x.\cos x\,dx$.

Integrating by parts regarding cos x as the 2nd function, we have

$I_n = \cos^{n-1}x.\sin x - \int (n-1)\cos^{n-2}x.(\sin x).\sin x\,dx$

$= \cos^{n-1}x.\sin x + (n+1)\int \cos^{n-2}x.\sin^2 x\,dx$

$= \cos^{n-1}x.\sin x + (n-1)\int \cos^{n-2}x\left(1-\cos^2 x\right)dx$

$= \cos^{n-1}x.\sin x + (n-1)\int \cos^{n-2}x\,dx - (n-1)\int \cos^n x\,dx$

$= \cos^{n-1}x \sin x + (n-1) I_{n-2} - (n-1) I_n$.

Transposing the last term to the left, we have

$I_n (1 + n - 1) = \cos^{n-1}x.\sin x + (n-1) I_{n-2}$

or $n I_n = \cos^{n-1}x.\sin x + (n-1) I_{n-2}$.

$$\therefore \int \cos^n dx = \frac{\cos^{n-1}x \sin x}{n} + \frac{n-1}{n}\int \cos^{n-2}x\,dx.$$

2.2. TO FIND REDUCTION FORMULA FOR $\int \tan^n x\,dx$ and $\int \cot^n x\,dx$.

(a) We have $\int \tan^n x\,dx = \int \tan^{n-2}x.\tan^2 x\,dx$ **(Note)**

$= \int \tan^{n-2}x.\left(\sec^2 x - 1\right)dx$

$= \int \tan^{n-2}x.\sec^2 x\,dx - \int \tan^{n-2}x\,dx$

$= \frac{(\tan x)^{n-2+1}}{n-2+1} - \int \tan^{n-2}x\,dx$

or $\int \tan^n x\,dx = \frac{\tan^{n-1}x}{n-1} - \int \tan^{n-2}x\,dx$,

which is the required reduction formula.

Application : *Evaluate* $\int \tan^4 x\,dx$.

Putting n = 4 in the above reduction formula, we have

$\int \tan^4 x\,dx = \frac{1}{3}\tan^3 x - \int \tan^2 x\,dx = \frac{1}{3}\tan^3 x - \int \left(\sec^2 x - 1\right)dx$

$= \frac{1}{3}\tan^3 x - \tan x + x$.

(b) We have $\int \cot^n x\,dx = \int \cot^{n-2} x.\cot^2 x\,dx$

$$= \int \cot^{n-2} x\left(\text{cosec}^2 x - 1\right)dx$$

$$= \int \cot^{n-2} x.\text{cosec}^2 x\,dx - \int \cot^{n-2} x\,dx$$

$$= -\frac{(\cot x)^{n-1}}{n-1} - \int \cot^{n-2} x\,dx$$

or $\cot^n x\,dx = -\dfrac{\cot^{n-1} x}{n-1} - \int \cot^{n-2} x\,dx,$

which is the required reduction formula.

Application: Putting n = 5 in the above reduction formula and applying it repeatedly, we have

$$\int \cot^5 x\,dx = -\frac{1}{4}\cot^4 x - \int \cot^3 x\,dx$$

$$= -\frac{1}{4}\cot^4 x - \left[-\frac{1}{2}\cot^2 x - \int \cot x\,dx\right]$$

$$= -\frac{1}{4}\cot^4 x + \frac{1}{2}\cot^2 x + \int \cot x\,dx$$

$$= -\frac{1}{4}\cot^4 x + \frac{1}{2}\cot^2 x + \log \sin x.$$

2.3. TO OBTAIN THE REDUCTION FORMULAE $\int \sec^n x\,dx$ and $\int \text{cosec}^n x\,dx$.

(a) We have $I_n = \int \sec^n x\,dx = \int \sec^{n-2} x.\sec^2 x\,dx$ **(Note)**

Integrating by parts regarding $\sec^2 x$ as the 2nd function, we have

$$I_n = \sec^{n-2} x \tan x - \int (n-2)\sec^{n-3} x \sec x \tan^2 x\,dx$$

$$= \sec^{n-2} x \tan x - (n-2)\int \sec^{n-2} x\left(\sec^2 x - 1\right)dx \quad \textbf{(Note)}$$

$$= \sec^{n-2} x \tan x - (n-2)\int \sec^n x\,dx + (n+2)\int \sec^{n-2} x\,dx.$$

Transposing the term containing $\int \sec^n x\,dx$ to the left, we have

$$(n-2+1)\int \sec^n x\,dx = \sec^{n-2} x \tan x + (n-2)\int \sec^{n-2} x\,dx$$

or $(n-1)\int \sec^n x\,dx = \sec^{n-2} x\tan x + (n-2)\int \sec^{n-2} x\,dx$.

Dividing both sides by (n – 1), we have

$$\int \sec^n x\,dx = \frac{\sec^{n-2} x\tan x}{n-1} + \frac{n-2}{n-1}\int \sec^{n-2} x\,dx,$$

which is the required reduction formula.

(b) To find the reduction formula for $\int \operatorname{cosec}^n x\,dx$, proceed exactly in the same way as in part (a). Thus, we get

$$\int \operatorname{cosec}^n x\,dx = -\frac{\operatorname{cosec}^{n-2} x\cot x}{n-1} + \frac{n-2}{n-1}\int \operatorname{cosec}^{n-2} x\,dx,$$

as in required reduction formula for $\int \operatorname{cosec}^n x\,dx$.

2.4. WALLI'S FORMULA

To evaluate $\int_0^{\pi/2} \sin^n x\,dx$ and $\int_0^{\pi/2} \cos^n x\,dx$

Proceeding as in the previous article, we have

$$\int \sin^n x\,dx = -\frac{\sin^{n-1} x\cos x}{n} + \frac{n-1}{n}\int \sin^{n-2} x\,dx.$$

$$\therefore \int_0^{\pi/2} \sin^n x\,dx = -\left[\frac{\sin^{n-1} x\cos x}{n}\right]_0^{\pi/2} + \frac{n-1}{n}\int_0^{\pi/2} \sin^{n-2} x\,dx$$

$$= 0 + \frac{n-1}{n}\int_0^{\pi/2} \sin^{n-2} x\,dx. \qquad \text{...(1)}$$

Putting (n – 2) in place of n in (1), we have

$$\int_0^{\pi/2} \sin^{n-2} x\,dx = \frac{n-3}{n-2}\int_0^{\pi/2} \sin^{n-4} x\,dx$$

Substituting this value in (1), we have

$$\int_0^{\pi/2} \sin^n x\,dx = \frac{n-1}{n}.\frac{n-3}{n-2}.\int_0^{\pi/2} \sin^{n-4} x\,dx$$

$$= \frac{n-1}{n}\cdot\frac{n-3}{n-2}\cdot\frac{n-5}{n-4}\cdot\int_0^{\pi/2} \sin^{n-6} x\,dx. \qquad \text{...(2)}$$

Now two cases arise viz., n is even or odd.

Case I: *When n is odd.* In this case by the repeated application of the reduction formula (1), the last integral of (2) is

$$\int_0^{\pi/2} \sin x\,dx = [-\cos x]_0^{\pi/2} = 1.$$

Hence when n is odd, from (2), we have

$$\int_0^{\pi/2} \sin^n x\,dx = \frac{n-1}{n}\cdot\frac{n-3}{n-2}\cdot\frac{n-5}{n-4}\cdots\frac{2}{3}\int_0^{\pi/2}\sin x\,dx$$

$$= \frac{n-1}{n}\cdot\frac{n-3}{n-2}\cdot\frac{n-5}{n-4}\cdots\frac{2}{3}\cdot 1 = \frac{(n-1)(n-3)\ldots 4.2}{n(n-2)\ldots 3.1}\cdot 1.$$

Case II: *When n is even.* In this case the last integral of (2) is

$$\int_0^{\pi/2}\sin^0 x\,dx = \int_0^{\pi/2} dx = [x]_0^{\pi/2} = \frac{\pi}{2}.$$

Hence when n is even, from (2), we have

$$\int_0^{\pi/2} \sin^n x\,dx = \frac{n-1}{n}\cdot\frac{n-3}{n-2}\cdot\frac{n-5}{n-4}\cdots\frac{3}{4}\cdot\frac{1}{2}\int_0^{\pi/2}\sin^0 x\,dx$$

$$= \frac{n-1}{n}\cdot\frac{n-3}{n-2}\cdot\frac{n-5}{n-4}\cdots\frac{3}{4}\cdot\frac{1}{2}\cdot\frac{\pi}{2} = \frac{(n-1)(n-3)\ldots 3.1}{n(n-2)\ldots 4.2}\cdot\frac{\pi}{2}.$$

If we evaluate $\int_0^{\pi/2}\cos^n x\,dx$, we get the same results.

$$\therefore \int_0^{\pi/2}\sin^n x\,dx = \int_0^{\pi/2}\cos^n x\,dx. \qquad \textbf{(Note)}$$

Note: Walli's formula is applicable only when the limits are from 0 to $\frac{1}{2}\pi$.

2.5. TO FIND A REDUCTION FORMULA FOR $\int \sin^m x \cos^n x\,dx$

Let $I_{m,n} = \int \sin^m x \cos^n x\,dx$

$$= \int \sin^m x \cos^{n-1} x \cos x\,dx = \int \cos^{n-1} x\cdot(\sin^m x\cos x)\,dx$$

Integrating by parts taking $\sin^m x \cos x$ as the second function, we get

$$I_{m,n} = \frac{\sin^{m+1} x}{m+1}\cos^{n-1} x + \frac{n-1}{m+1}\int \sin^{m+1} x \cos^{n-2} x \sin x\,dx$$

$$= \frac{\sin^{m+1} x}{m+1}\cos^{n-1} x + \frac{n-1}{m+1}\int \sin^m x \cos^{n-2} x \sin^2 x\,dx$$

$$= \frac{\sin^{m+1} x}{m+1}\cdot\cos^{n-1} x + \frac{n-1}{m+1}\int \sin^m x \cos^{n-2} x\cdot(1-\cos^2 x)\,dx$$

$$= \frac{\sin^{m+1} x \cos^{n-1} x}{m+1} + \frac{n-1}{m+1}\int \sin^m x \cos^{n-2} x\, dx - \frac{n-1}{m+1} I_{m, n}$$

Transposing the last term to the left, we have

$$I_{m, n}\left(1 + \frac{n-1}{m+1}\right) = \frac{\sin^{m+1} x \cdot \cos^{n-1} x}{m+1} + \frac{n-1}{m+1} I_{m, n-2}$$

or $$I_{m, n}\left(1 + \frac{n-1}{m+1}\right) = \frac{\sin^{m+1} x \cos^{n-1} x}{m+1} + \frac{n-1}{m+1} I_{m, n-2}$$

$$I_{m,n}\left(\frac{m+n}{m+1}\right) = \frac{\sin^{m+1} x \cos^{n-1} x}{m+1} + \frac{n-1}{m+1} I_{m, n-2}$$

Thus the required reduction formula is

$$I_{m, n} = \frac{\sin^{m+1} x \cdot \cos^{n-1} x}{m+n} + \frac{(n-1) I_{m, n-2}}{m+n}$$

Note: If we write $I_{m, n} = \int \sin^m x \cos^n x\, dx$

$$= \int \sin^{m-1} x \cdot (\cos^n x \sin x)\, dx$$

Then integrating by parts regarding $\cos^n x \sin x$ as the 2nd function, the reduction formula can be obtained as

$$I_{m, n} = -\frac{\sin^{m-1} x \cdot \cos^{n+1} x}{m+n} + \frac{m-1}{m+n} I_{m-2, n}$$

Similarly other four reduction formulae for $\int \sin^m x \cos^n x\, dx$ may be obtained as

$$I_{m, n} = -\frac{\sin^{m+1} x \cos^{n+1} x}{n+1} + \frac{m+n+2}{n+1} I_{m, n+2}$$

[To obtain this reduction formula put (n + 2) in place of n in the reduction formula obtained.]

$$I_{m, n} = \frac{\sin^{m+1} x \cos^{n+1} x}{m+1} + \frac{m+n+2}{m+1} I_{m+2, n}$$

$$I_{m, n} = -\frac{\sin^{m+1} x \cos^{n+1} x}{m+1} + \frac{n-1}{m+1} I_{m-2, n+2}$$

$$I_{m, n} = \frac{\sin^{m+1} x \cos^{n-1} x}{m+1} + \frac{n-1}{m+1} I_{m+2, n-2}.$$

2.6. GAMMA FUNCTION

The definite integral $\int_0^{\infty} e^{-x}x^{n-1}dx$ is called the *second Eulerian integral* and is denoted by the symbol Γ (n) [read as Gamma n].

Properties of Gamma Function (Commit to Memory)

$\Gamma(n+1) = n\Gamma n;\ \Gamma 1 = 1;\ \Gamma\frac{1}{2} = \sqrt{\pi}$

$\Gamma(n) = (n-1)!$ provided n is a positive integer. Thus $\Gamma(10) = 9!$.

Also $\Gamma\frac{9}{2} = \frac{7}{2}\Gamma\frac{7}{2} = \frac{7}{2}\cdot\frac{5}{2}\Gamma\frac{5}{2} = \frac{7}{2}\cdot\frac{5}{2}\cdot\frac{3}{2}\Gamma\frac{3}{2}$

$= \frac{7}{2}\cdot\frac{5}{2}\cdot\frac{3}{2}\cdot\frac{1}{2}\Gamma\frac{1}{2} = \frac{7}{2}\cdot\frac{5}{2}\cdot\frac{3}{2}\cdot\frac{1}{2}\sqrt{\pi} = \frac{105}{16}\sqrt{\pi}$

2.7. TO SHOW THAT

$$\int_0^{\pi/2} \sin^m x \cos^n x\, dx = \frac{\Gamma\left(\frac{m+1}{2}\right)\Gamma\left(\frac{n+1}{2}\right)}{2\Gamma\left(\frac{m+n+2}{2}\right)}$$

where m and n are psoitve integers.
We have

$$\int \sin^m x \cos^n x\, dx = \frac{\sin^{m+1} x \cos^{n-1} x}{m+n} + \frac{n-1}{m+n}\int \sin^m x \cos^{n-2} x\, dx$$

$$\therefore \int_0^{\pi/2} \sin^m x \cos^n x\, dx$$

$$= \left[\frac{\sin^{m+1} x \cos^{n-1} x}{m+n}\right]_0^{\pi/2} + \frac{n-1}{m+n}\int_0^{\pi/2} \sin^m x \cos^{n-2} x\, dx$$

$$= 0 + \frac{n-1}{m+n}\int_0^{\pi/2} \sin^m x \cos^{n-2} x\, dx$$

i.e., $$\int_0^{\pi/2} \sin^m x \cos^n x\, dx = \frac{(n-1)}{(m+n)}\int_0^{\pi/2} \sin^m x \cos^{n-2} x\, dx \qquad ...(1)$$

Now four cases arise according as m and n take different types of values, odd or even.

Case I. *When m and n are both even.*

Successively applying the formula (1) till the power of cos x becomes zero, we have

$$\int_0^{\pi/2} \sin^m x \cdot \cos^n x\,dx$$

$$= \frac{(n-1)}{(m+n)} \cdot \frac{(n-3)}{(m+n-2)} \cdot \frac{(n-5)}{(m+n-4)} \cdots \frac{1}{(m+2)} \int_0^{\pi/2} \sin^m x\,dx$$

Also $\int_0^{\pi/2} \sin^m x\,dx = \frac{m-1}{m} \cdot \frac{m-3}{m-2} \cdot \frac{m-5}{m-4} \cdots \frac{1}{2} \cdot \frac{\pi}{2}$. [See 5.2]

Therefore $\int_0^{\pi/2} \sin^m x \cos^n x\,dx$

$$= \frac{(n-1)(n-3)(n-5)\ldots 1}{(m+n)(m+n-2)(m+n-4)\ldots(m+2)} \times \frac{(m-1)(m-3)(m-5)\ldots 3.1}{m(m-2)(m-4)\ldots 4.2} \cdot \frac{\pi}{2}$$

$$= \frac{\left\{\left(\frac{n-1}{2}\right)\left(\frac{n-3}{2}\right)\left(\frac{n-5}{2}\right)\cdots\frac{1}{2}\right\}\left\{\left(\frac{m-1}{2}\right)\left(\frac{m-3}{2}\right)\cdots\frac{1}{2}\right\}}{\left\{\left(\frac{m+n}{2}\right)\left(\frac{m+n-2}{2}\right)\cdots\frac{4}{2} \cdot \frac{2}{2}\right\}} \cdot \frac{\pi}{2}$$

$$= \frac{\Gamma\left(\frac{m+1}{2}\right)\Gamma\left(\frac{n+1}{2}\right)}{2\Gamma\left(\frac{m+n+2}{2}\right)}$$

Similarly, the cases for other values of m and n may be considered and it may be verified that the result is true in other cases too.

Thus for all positive integral values of m and n, we have

$$\int_0^{\pi/2} \sin^m x \cos^n x\,dx = \frac{\Gamma\left(\frac{m+1}{2}\right) \cdot \Gamma\left(\frac{n+1}{2}\right)}{2\Gamma\left(\frac{m+n+2}{2}\right)} \qquad \text{(Remember)}$$

Walli's Formula : [An easy way to evaluate $\int_0^{\pi/2} \sin^m x \cos^n x\,dx$ when *m* and *n* are + *ve* integers]. We have $\int_0^{\pi/2} \sin^m x \cos^n x\,dx$.

$$= \frac{(m-1)(m-3)(m-5)\ldots(n-1)(n-3)(n-5)\ldots}{(m+n)(m+n-2)(m+n-4)\ldots} \times k$$

where k is $\frac{1}{2}\pi$ if m and n are both even, otherwise k = 1. The last factor in each of the three products is either 1 or 2. In case any of m or n is 1, we simply write 1 as the only factor to replace its product. This formula is equally applicable if any of m or n is zero provided we put 1 as the only factor in its product and we regard 0 as even.

2.8. INTEGRATION OF x^n SIN mx AND x^n COS mx

(a) $\int x^n \sin mx\, dx$: To form the reduction formula, integrating by parts regarding sin mx as the 2nd function, we have

$$\int x^n \sin mx\, dx = -\frac{x^n \cos mx}{m} + \int nx^{n-1} \frac{\cos mx\, dx}{m}$$

$$= -\frac{x^n \cos mx}{m} + \frac{n}{m}\left[\frac{x^{n-1}\sin mx}{m} - \frac{n-1}{m}\int x^{n-2} \sin mx\, dx\right]$$

[again integrating by parts regarding cos mx as the 2nd function]

$$= -\frac{x^n \cos mx}{m} + \frac{nx^{n-1}}{m^2}\sin mx - \frac{n(n-1)}{m^2}\int x^{n-2} \sin mx\, dx$$

Above is the required reduction formula. Successively applying this formula we are left with $\int x \sin mx\, dx$ or $\int \sin mx\, dx$ according as n is odd or even.

(b) $\int x^n \cos mx\, dx$: Integrating by parts regarding cos mx as the 2nd function, we have

$$\int x^n \cos mx\, dx = \frac{x^n \sin mx}{m} - \int \frac{nx^{n-1} \sin mx\, dx}{m}$$

$$= \frac{x^n \sin mx}{m} - \frac{n}{m}\int x^{n-1} \sin mx\, dx$$

$$= \frac{x^n \sin mx}{m} - \frac{n}{m}\left[x^{n-1}\cdot\left(\frac{\cos mx}{m}\right) - \frac{n-1}{m}\int x^{n-2} \cos mx\, dx\right]$$

[again integrating by parts regarding sin mx as 2nd function]

$$= \frac{x^n \sin mx}{m} + \frac{nx^{n-1}\cos mx}{m^2} - \frac{n(n-1)}{m^2}\int x^{n-2} \cos mx\, dx$$

which is the required reduction formula.

2.9. REDUCTION FORMULAE FOR

$\int x \sin^n x\, dx$ and $\int x \cos^n x\, dx$

Let $I_n = \int x \sin^n x\, dx = \int (x \sin^{n-1} x) \cdot \sin x\, dx$ (Note)

$$= (x \sin^{n-1} x) \cdot (-\cos x) + \int \cos x\, [\sin^{n-1} x + x(n-1) \sin^{n-2} x \cos x]\, dx$$

integrating by parts regarding sin x as 2nd function

$$= -x \sin^{n-1} x \cos x + \int \sin^{n-1} x \cos x\, dx + (n-1) \int x \sin^{n-2} x \cos^2 x\, dx$$

$$= -x \sin^{n-1} x \cos x + \frac{1}{n} \sin^n x + (n-1) \int x \sin^{n-2} x \cdot (1 - \sin^2 x)\, dx$$

$$= -x \sin^{n-1} x \cos x + \frac{1}{n} \sin^n x + (n-1) \int x \sin^{n-2} x\, dx - (n-1) \int x \sin^n x\, dx$$

$$= -x \sin^{n-1} x \cos x + \frac{1}{n} \sin^n x + (n-1) I_{n-2} - (n-1) I_n$$

Transposing the last term to the left, we have

$$I_n (1 + n - 1) = -x \sin^{n-1} x \cos x + \frac{1}{n} \sin^n x + (n-1) I_{n-2}$$

or $\quad nI_n = -x \sin^{n-1} x \cos x + \frac{1}{n} \sin^n x + (n-1) I_{n-2}$

or $\quad I_n = -\frac{x}{n} \cdot \sin^{n-1} x \cdot \cos x + \frac{\sin^n x}{n^2} + \frac{(n-1)}{n} I_{n-2}.$

which is the required reduction formula.

Similarly, reduction formula for $\int x \cos^n x\, dx$ is

$$I_n = \int x \cos^n x\, dx = \frac{x \cos^{n-1} x \sin x}{n} + \frac{\cos x}{n^2} + \frac{(n-1)}{n} I_{n-2}.$$

2.10. REDUCTION FORMULAE FOR

$\int e^{ax} \sin^n bx\, dx$ and $\int e^{ax} \cos^n bx\, dx$

(a) Let $I_n = \int e^{ax} \sin^n bx\, dx$

$$= \frac{e^{ax}}{a} \sin^n bx - \frac{nb}{a} \int e^{ax} \sin^{n-1} bx \cos bx\, dx \qquad ...(1)$$

integrating by parts taking e^{ax} as the 2nd function.

Now $\int e^{ax} \sin^{n-1} bx \cos bx\, dx = \frac{e^{ax}}{a} (\sin^{n-1} bx \cos bx)$

$$- \int \frac{e^{ax}}{a} [(n-1) b \sin^{n-2} bx \cos^2 bx - b \sin^n bx]\, dx$$

integrating by parts taking e^{ax} as the 2nd function

$$= \frac{e^{ax}}{a} (\sin^{n-1} bx \cos bx)$$

$$- \frac{b}{a} \int e^{ax} [(n-1) \sin^{n-2} bx (1 - \sin^2 bx) - \sin^n bx]\, dx$$

$$= \frac{e^{ax}}{a} (\sin^{n-1} bx \cos bx)$$

$$- \frac{b}{a} \int e^{ax} [(n-1) \sin^{n-2} bx - n \sin^n bx] dx$$

$$= \frac{e^{ax}}{a} (\sin^{n-1} bx \cos bx) - (n-1) \frac{b}{a} \int e^{ax} \sin^{n-2} bx\, dx + \frac{nb}{a} I_n$$

Substituting this value in (1), we get

$$I_n = \frac{e^{ax}}{a} \sin^n bx - \frac{nb}{a^2} e^{ax} \sin^{n-1} bx \cos bx$$

$$+ n(n-1) \frac{b^2}{a^2} \int e^{ax} \sin^{n-2} bx\, dx - n^2 \frac{b^2}{a^2} I_n$$

Transposing the last term to L.H.S., we get

$$\left(1 + \frac{n^2 b^2}{a^2}\right) I_n = \frac{e^{ax}}{a^2} (a \sin bx - nb \cos bx) \sin^{n-1} bx$$

$$+ n(n-1) \frac{b^2}{a^2} I_{n-2}$$

$$\therefore\ I_n = \frac{e^{ax}}{a^2 + n^2 b^2} (a \sin^n bx - nb \sin^{n-1} bx \cos bx)$$

$$+ \frac{n(n-1)}{a^2 + n^2 b^2} I_{n-2}$$

which is the required reduction formula.

Similarly, $\int e^{ax} \cos^n bx\, dx$

$$= \frac{e^{ax}}{a^2 + n^2b^2}(a \cos^n bx + nb \sin bx \cos^{n-1} bx) + \frac{n(n-1)}{a^2 + n^2b^2} I_{n-2}$$

Note: The above formulae should not be applied when n is small. In that case $\sin^n bx$ and $\cos^n bx$ are converted in terms of multiples of angles.

2.11. REDUCTION FORMULAE FOR $\int x^n e^{ax} \sin bx\, dx$ and $\int x^n e^{ax} \cos bx\, dx$

We know that $\int e^{ax} \sin bx\, dx = \frac{e^{ax}}{r} \sin(bx - \phi)$

where $r = \sqrt{(a^2 + b^2)}$ and $\phi = \tan^{-1}\left(\frac{b}{a}\right)$

Now $\frac{1}{r}\int e^{ax} \sin(bx - \phi)\, dx = \frac{1}{r}\left[\frac{1}{r} e^{ax} \sin\{(bx - \phi) - \phi\}\right]$

$$= \frac{1}{r^2}\int e^{ax} \sin(bx - 2\phi)$$

Similarly $\frac{1}{r^2}\int e^{ax} \sin(bx - 2\phi)\, dx$

$$= \frac{1}{r^3} e^{ax} \sin(bx - 3\phi), \text{ and so on.}$$

Now $\int x^n e^{ax} \sin bx\, dx$ can be easily evaluated by repeatedly integrating by parts taking function of the type $e^{ax} \sin bx$ as the 2nd function.

Similarly we can obtain a reduction formula for

$\int x^n e^{ax} \cos bx\, dx$.

2.12. REDUCTION FORMULA FOR $\int \cos^m x \sin nx\, dx$

Let $I_{m,n} = \int \cos^m x \sin nx\, dx$. Integrating by parts regarding sin nx as the 2nd function, we have

$$I_{m,n} = \frac{\cos^m x\,(-\cos nx)}{n} - \int \frac{m\cos^{m-1} x \cdot (-\sin x)\,(-\cos nx)\,dx}{n}$$

$$= \frac{-\cos^m x \cos nx}{n} - \frac{m}{n}\int \cos^{m-1} x \cos nx \cdot \sin x\,dx \qquad ...(1)$$

Now $\sin\{(n-1)x\} = \sin(nx - x) = \sin nx \cos x - \cos nx \sin x.$

$\therefore\ \cos nx \sin x = \sin nx \cos x - \sin(n-1)x.$

$$\therefore\ I_{m,n} = -\frac{\cos^m x \cos nx}{n} - \frac{m}{n}\int \cos^{m-1} x\,\{\sin nx \cos x - \sin(n-1)x\}\,dx$$

$$= -\frac{\cos^m x \cos nx}{n} - \frac{m}{n}\int \cos^m x \sin nx\,dx + \frac{m}{n}\int \cos^{m-1} x \sin(n-1)x\,dx$$

Transposing the middle term to the left and simplifying, we get

$$I_{m,n} = -\frac{\cos^m x \cos nx}{m+n} + \frac{m}{m+n}\int \cos^{m-1} x \sin(n-1)x\,dx$$

which is the required reduction formula.

Deduction : If in the above integral we take the limits of integration as 0 to $\frac{1}{2}\pi$, we find that

$$\int_0^{\pi/2} \cos^m x \sin nx\,dx = \frac{1}{m+n} + \frac{m}{m+n}\int_0^{\pi/2} \cos^{m-1} x \sin(n-1)x\,dx$$

2.13. REDUCTION FORMULA FOR $\int \cos^m x \cos nx\,dx$

Let $I_{m,n} = \int \cos^m x \cdot \cos nx\,dx$

$$= \cos^m x \cdot \left(\frac{\sin nx}{n}\right) - \int m\cos^{m-1} x \cdot (-\sin x)\left(\frac{\sin nx}{n}\right)dx$$

integrating by parts taking cos nx as the 2nd function

$$= \frac{\cos^m x \sin nx}{n} + \frac{m}{n}\int \cos^{m-1} x \cdot \sin nx \sin x\,dx$$

But $\cos(n-1)x = \cos nx \cos x + \sin nx \sin x$

$\therefore \sin nx \sin x = \cos(n-1)x - \cos nx \cos x$

Hence $I_{m,n} = \dfrac{\cos^m x \cdot \sin nx}{n}$

$$+\frac{m}{n}\int \cos^{m-1} x \{\cos(n-1)x - \cos nx \cos x\}\, dx$$

$$= \frac{\cos^m x \sin nx}{n} + \frac{m}{n}\int \cos^{m-1} x \cos(n-1)x\, dx$$

$$\frac{-m}{n}\int \cos^m x \cos nx\, dx$$

$$= \frac{\cos^m x \sin nx}{n} + \frac{m}{n} I_{m-1,n-1} - \frac{m}{n} I_{m,n}$$

Transposing the last term to the left, we have

$$\left(1+\frac{m}{n}\right) I_{m,n} = \frac{\cos^m x \sin nx}{n} + \frac{m}{n} I_{m-1,n-1}$$

or $I_{m,n} = \dfrac{\cos^m x \sin nx}{m+n} + \dfrac{m}{m+n} I_{m-1,n-1}$,

which is the reqwuired reduction formula.

Deduction : $\int_0^{\pi/2} \cos^m x \cos nx\, dx$

$$= \left[\frac{\cos^m x \sin nx}{m+n}\right]_0^{\pi/2} + \frac{m}{(m+n)}\int_0^{\pi/2} \cos^{m-1} x \cos(n-1)x\, dx$$

$$= 0 + \frac{m}{(m+n)}\int_0^{\pi/2} \cos^{m-1} x \cos(n-1)x\, dx$$

or $I_{m,n} = \displaystyle\int_0^{\pi/2} \cos^m x \cos nx\, dx = \frac{m}{m+n} I_{m-1,n-1}$

MISCELLANEOUS EXAMPLES

Example 1:

If $I_n = \displaystyle\int_0^{\pi/2} x^n \sin(2p+1)x\, dx$, *prove that*

$$I_n + \frac{n(n-1)}{(p+1)^2} I_{n-2} = (-1)^p - \frac{n}{(2p+1)^2}\left(\frac{\pi}{2}\right)^{n-1}$$

where n and p are positive integers.

Hence deduce that $\int_0^{\pi/2} x^3 \sin 3x\, dx = \frac{2}{27} - \frac{\pi^2}{12}$.

Solution:

We have $I_n = \int_0^{\pi/2} x^n \sin(2p+1)x\, dx$

$$= \left[\left\{-x^n \frac{\cos(2p+1)x}{2p+1}\right\}_0^{\pi/2} + \frac{n}{(2p+1)} \int_0^{\pi/2} x^{n-1} \cos(2p+1)x\, dx\right]$$

integrating by parts taking sin (2p + 1) x as the 2nd function

$$= 0 + \frac{n}{(2p+1)}\left[x^{n-1} \cdot \frac{\sin(2p+1)x}{(2p+1)}\right]_0^{\pi/2} - \frac{n(n-1)}{(2p+1)^2} I_{n-2}$$

again integrating by parts

$$= \frac{n}{(2p+1)^2}\left(\frac{\pi}{2}\right)^{n-1} \sin(2p+1)\frac{\pi}{2} - \frac{n(n-1)}{(2p+1)^2} I_{n-2}$$

$$\therefore I_n + \frac{n(n-1)}{(2p+1)^2} I_{n-2} = (-1)^p \frac{n}{(2p+1)^2}\left(\frac{\pi}{2}\right)^{n-1} \quad ...(1)$$

Now to evaluate $\int_0^{\pi/2} x^3 \sin 3x\, dx$,

put n = 3, p = 1 in (1).

$$\therefore I_3 = -1 \cdot \frac{3}{3^2}\left(\frac{\pi}{2}\right)^{3-1} - \frac{3 \cdot 2}{3^2} I_1 = \frac{\pi^2}{12} - \frac{2}{3}\int_0^{\pi/2} x \sin 3x\, dx$$

$$= -\frac{\pi^2}{12} - \frac{2}{3}\left[\left\{-x \frac{\cos 3x}{3}\right\}_0^{\pi/2} + \frac{1}{3}\int_0^{\pi/2} \cos 3x\, dx\right]$$

$$= -\frac{\pi^2}{12} - \frac{2}{3^3}[\sin 3x]_0^{\pi/2} = -\frac{\pi^2}{12} + \frac{2}{27} = \frac{2}{27} - \frac{\pi^2}{12}.$$

Example 2:

Evaluate $\int (\sin^{-1} x)^n\, dx$.

Solution:

Do your self.

Put $\sin^{-1} x = \theta$

i.e., $x = \sin\theta$ and $dx = \cos\theta\, d\theta$.

Then $\int (\sin^{-1} x)^n\, dx = \int \theta^n \cos\theta\, d\theta$.

Example 3:

Evaluate $\int (\cos^{-1} x)^n\, dx$

Solution:

Do your self.

Here put $\cos^{-1} x = \theta$

i.e., $x = \cos\theta$ and $dx = -\sin\theta\, d\theta$.

Then $\int (\cos^{-1} x)^n\, dx = -\int \theta^n \sin\theta\, d\theta$.

Example 4:

Evaluate $\int_0^{\pi} x \sin^3 x\, dx$

Solution:

We have

$$\int x \sin^n x\, dx = -\frac{x}{n} \cdot \sin^{n-1} x \cos x + \frac{\sin^n x}{n^2} + \frac{(n-1)}{n} \int x \sin^{n-2} x\, dx \quad ...(1)$$

Putting n = 3 in (1), we have

$$\int x \sin^3 x\, dx = -\frac{1}{3} x \sin^2 x \cos x + \frac{1}{9} \sin^3 x + \frac{2}{3} \int x \sin x\, dx$$

Now $\int x \sin x\, dx = x \cdot (-\cos x) - \int 1 (1 - \cos x)\, dx$

$= -x \cos x + \sin x$.

$$\therefore \int_0^{\pi} x \sin^3 x\, dx = \left[-\frac{1}{3} x \sin^2 x \cos x + \frac{1}{9} \sin^3 x - \frac{2}{3} x \cos x + \frac{2}{3} \sin x \right]_0^{\pi}$$

$$= -\frac{2}{3} \pi \cos\pi = -\frac{2}{3} \pi (-1) = \frac{2}{3} \pi.$$

Example 5:

Evaluate $\int x \sin^4 x\, dx$.

Solution:

Here putting n = 4 in the reduction formula (1) and proceeding as in part (a), we have

$$\int x \sin^4 x\, dx$$

$$= -\frac{x \sin^3 x \cos x}{4} + \frac{\sin^4 x}{16} + \frac{3}{16}x^2 - \frac{3}{16} x \sin 2x - \frac{3}{32}\cos 2x.$$

Example 6:

Evaluate $\int_0^{\pi/2} x \cos^3 x\, dx$

Solution:

We know that

$$\int x \cos^n x\, dx = \frac{x \cos^{n-2} x \sin x}{n} + \frac{\cos^n x}{n^2} + \frac{n-1}{n} I_{n-2},$$

[Derive it here]

Putting n = 3, we have $\int x \cos^3 x\, dx$

$$= \frac{1}{3} x \cos^2 x \sin x + \frac{1}{9}\cos^3 x + \frac{2}{3}\int x \cos x\, dx$$

Now $\int x \cos x\, dx = x \cdot \sin x - \int 1 \sin x\, dx = x \sin x + \cos x$

$$\therefore \int_0^{\pi/2} x \cos^3 x\, dx = \left[\frac{1}{3} x \cos^2 x \sin x + \frac{1}{9}\cos^3 x + \frac{2}{3}(x \sin x + \cos x)\right]_0^{\pi/2}$$

$$= \frac{1}{3}\pi - \frac{1}{9} - \frac{2}{3} = \frac{1}{3}\pi - \frac{7}{9} = \frac{1}{3}\left[\pi - \frac{7}{3}\right].$$

Example 7:

If $u_n = \int_0^{\pi/2} \theta \sin^n \theta\, d\theta$ *and* $n > 1$, *prove that*

$$u_n = \frac{(n-1)}{n} u_{n-2} + \frac{1}{n^2}.$$

Hence deduce that $u_5 = \dfrac{149}{225}$.

Solution:

We know that

$$\int \theta \sin^n \theta \, d\theta = -\frac{\theta \sin^{n-1}\theta \cos\theta}{n} + \frac{\sin^n \theta}{n^2} + \frac{(n-1)}{n}\int \theta \sin^{n-2}\theta \, d\theta. \qquad \text{[Derive it here]}$$

$$\therefore\ u_n = \int_0^{\pi/2} \theta \sin^n \theta \, d\theta = \left[-\frac{\theta \sin^{n-1}\theta \cos\theta}{n} + \frac{\sin^n\theta}{n^2}\right]_0^{\pi/2} + \frac{(n-1)}{n}\int_0^{\pi/2} \theta \sin^{n-2}\theta \, d\theta$$

$$= \left[\frac{1}{n^2} - 0\right] + \frac{(n-1)}{n} u_{n-2} = \frac{1}{n^2} + \frac{(n-1)}{n} u_{n-2} \qquad ...(1)$$

Putting n = 5 in the above reduction formula, we have

$$u_5 = \frac{1}{25} + \frac{4}{5} u_3 = \frac{1}{25} + \frac{4}{5}\left[\frac{1}{9} + \frac{2}{3} u_1\right], \qquad \text{[Putting n = 3 in (1)]}$$

$$= \frac{1}{25} + \frac{4}{45} + \frac{8}{15} u_1 = \left(\frac{1}{25} + \frac{4}{45}\right) + \frac{8}{15}\int_0^{\pi/2} \theta \sin\theta \, d\theta$$

$$= \left(\frac{29}{225}\right) + \frac{8}{15}\left[(-\theta\cos\theta)_0^{\pi/2} + \int_0^{\pi/2} \cos\theta \, d\theta\right]$$

$$= \frac{29}{225} + \frac{8}{15}[0 + \sin\theta]_0^{\pi/2} = \frac{29}{225} + \frac{8}{15} = \frac{149}{225}.$$

Example 8:

Integrating by parts twice or otherwise, obtain a reduction formula for

$$I_m = \int_0^{\infty} e^{-x} \sin^m x \, dx, \text{ where } m \geq 2$$

in the form $(1 + m^2) I_m = m(m-1) I_{m-2}$ *and hence evaluate* I_4.

Solution:

We have $I_m = \int_0^{\infty} e^{-x} \sin^m x \, dx$

$$= \left[\sin^m x \cdot (-e^{-x})\right]_0^{\infty} + \int_0^{\infty} m \sin^{m-1} x \cos x \, e^{-x} dx,$$

integrating by parts taking e^{-x} as the second function

$$= 0 + m\int_0^{\infty} (\sin^{m-1} x \cos x) \cdot e^{-x} dx$$

$$= m\left[\sin^{m-1} x \cos x \cdot (-e^{-x})\right]_0^{\infty}$$

$$- m\int_0^{\infty} [-\sin^m x + (m-1)\sin^{m-2} x \cos^2 x] \cdot (-e^{-x})\, dx$$

$$= 0 + m\int_0^{\infty} e^{-x}[-\sin^m x + (m-1)\sin^{m-2} x (1-\sin^2 x)]\, dx$$

$$= m\int_0^{\infty} e^{-x}[(m-1)\sin^{m-2} x - m\sin^m x]\, dx$$

$$= m(m-1)I_{m-2} - m^2 I_m$$

or $(1 + m^2)\, I_m = m\,(m-1)\, I_{m-2}.$

or $I_m = \dfrac{m(m-1)}{1+m^2} I_{m-2}$

To evaluate I_4, putting m = 4 in (1), we get

$$I_4 = \frac{4(4-1)}{1+16} I_2 = \frac{12}{17} I_2$$

$$= \frac{12}{17}\left[\frac{2(2-1)}{1+4} I_0\right] = \frac{24}{85} I_0, \qquad \text{[To get } I_2\text{, we put m = 2 in (1)]}$$

$$= \frac{24}{85}\int_0^{\infty} e^{-x}\sin^0 x\, dx = \frac{24}{85}\int_0^{\infty} e^{-x} dx = -\frac{24}{85}\left[e^{-x}\right]_0^{\infty} = \frac{24}{85}$$

Example 9:

Evaluate $\int e^x (x\cos x + \sin x)\, dx$.

Solution:

The given integral $= \int xe^x \cos x\, dx + \int e^x \sin x\, dx = I_1 + I_2$, say

Now $I_1 = \int x \cdot e^x \cos x\, dx = x \cdot \frac{1}{2} e^x (\cos x + \sin x)$

$$- \int \frac{1}{2} e^x (\cos x + \sin x)\, dx,$$

integrating by parts taking ($e^x \cos x$) as 2nd function

$$= \frac{1}{2} xe^x (\cos x + \sin x) - \frac{1}{2}\int e^x \cos x\, dx - \frac{1}{2}\int e^x \sin x\, dx$$

$$= \frac{1}{2}xe^x(\cos x + \sin x) - \frac{1}{2}\cdot\frac{1}{2}e^x(\cos x + \sin x)$$

$$-\frac{1}{2}\cdot\frac{1}{2}e^x(\sin x - \cos x)$$

$$= \frac{1}{2}xe^x(\cos x + \sin x) - \frac{1}{2}e^x \sin x$$

And $I_2 = \int e^x \sin x\, dx = \frac{1}{2}e^x(\sin x - \cos x)$

$\therefore$ required integral $= I_1 + I_2$

$$= \frac{1}{2}x\, e^x(\cos x + \sin x) - \frac{1}{2}e^x \sin x + \frac{1}{2}e^x(\sin x - \cos x)$$

$$\frac{1}{2}e^x[x(\cos x + \sin x) - \cos x].$$

Remember:

$$\int e^x \cos x\, dx = \frac{1}{2}e^x(\cos x + \sin x)$$

and $\int e^x \sin x\, dx = \frac{1}{2}e^x(\sin x - \cos x)$

Example 10:

Evaluate $\int_0^\infty x e^{-2x} \cos x\, dx$

Solution:

The given integral $I = \int_0^\infty x\cdot(e^{-2x}\cos x)\, dx$

Integrating by parts taking $e^{-2x}\cos x$ as the 2nd function, we have

$$I = \left[x\frac{e^{-2x}}{r}\cos(x-\phi)\right]_0^\infty - \int_0^\infty 1\cdot\frac{e^{-2x}}{r}\cos(x-\phi)\, dx$$

where $\phi = \tan^{-1}\left(\frac{b}{a}\right) = \tan^{-1}\left(-\frac{1}{2}\right)$

and $r = \sqrt{(a^2+b^2)} = \sqrt{(4+1)} = \sqrt{5}$

$$= \left[\lim_{x\to\infty}\frac{1}{r}\frac{x}{e^{2x}}\cos(x-\phi) - 0\right] - \frac{1}{r}\int_0^\infty e^{-2x}\cos(x-\phi)\, dx$$

$$= 0 - \frac{1}{\sqrt{5}}\int_0^\infty e^{-2x}\cos(x-\phi)\, dx$$

$$= -\frac{1}{\sqrt{5}}\left[\frac{1}{\sqrt{5}}e^{-2x}\cos(x-2\phi)\right]_0^\infty$$

$$= -\left(\frac{1}{5}\right)[0 - e^0\cos(-2\phi)] = \frac{1}{5}\cos 2\phi$$

where $\phi = \tan^{-1}\left(\frac{b}{a}\right) = \tan^{-1}\left(-\frac{1}{2}\right)$

$$= \frac{1}{5}\left[\frac{\{1-\tan^2\phi\}}{\{1+\tan^2\phi\}}\right], \text{ where } \tan\phi = -\frac{1}{2}$$

$$= \frac{1}{5}\left[\frac{\left(1-\frac{1}{4}\right)}{\left(1+\frac{1}{4}\right)}\right] = \frac{1}{5}\cdot\frac{3}{5} = \frac{3}{25}.$$

Example 11:

Evaluate $\int x^2 e^{2x\cos\alpha}\sin(2x\sin\alpha)\,dx$

Solution:

We know that

$$\int e^{ax}\sin bx\,dx = \frac{1}{r}e^{ax}\sin(bx-\phi)$$

where $r = \sqrt{(a^2+b^2)}$ and $\phi = \tan^{-1}\left(\frac{b}{a}\right)$

If we take $a = 2\cos\alpha$

and $b = 2\sin\alpha$, we have

$$r = \sqrt{(a^2+b^2)} = 2 \text{ and } \phi = \tan^{-1}\left(\frac{b}{a}\right) = \tan^{-1}(\tan\alpha) = \alpha$$

Now the given integral $= \int x^2\cdot e^{ax}\sin bx\,dx$

$$= x^2\left[\frac{e^{ax}}{r}\sin(bx-\phi)\right] - \int 2x\cdot\left[\frac{e^{ax}}{r}\sin(bx-\phi)\right]dx,$$

integrating by parts taking ($e^{ax}\sin bx$) as 2nd function

$$= \frac{1}{2}x^2\cdot e^{ax}\sin(bx-\phi) - \int x\,e^{ax}\sin(bx-\phi)\,dx \quad ...(1)$$

$[\because\ r = 2]$

Also $\int x \cdot e^{ax} \sin(bx - \phi)\, dx$

$$= \left[x \cdot \left(\frac{1}{r}\right) \cdot e^{ax} \sin(bx - \phi) - \int \left(\frac{1}{r}\right) \cdot e^{ax} \sin(bx - 2\phi)\, dx\right],$$

again integrating by parts

$$= \frac{1}{2} xe^{ax} \sin(bx - 2\phi) - \frac{1}{2} \int e^{ax} \sin(bx - 2\phi), \qquad [\because r = 2]$$

$$= \frac{1}{2} xe^{ax} \sin(bx - 2\phi) - \frac{1}{2} \cdot \frac{1}{2} e^{ax} \sin(bx - 3\phi) \qquad \text{...(2)}$$

Hence from (1) and (2), the given integral

$$= \frac{1}{2} x^2 e^{ax} \sin(bx - \phi) - \frac{1}{2} xe^{ax} \sin(bx - 2\phi) + \frac{1}{4} e^{ax} \sin(bx - 3\phi)$$

where $a = 2 \cos \alpha$, $b = 2 \sin \alpha$ and $\phi = \alpha$.

Example 12:

Prove that $\int_0^{\pi/2} \cos^m x \sin mx\, dx$

$$= \frac{1}{2^{m+1}} \left[2 + \frac{2^2}{2} + \frac{2^3}{3} + \frac{2^4}{4} + \ldots + \frac{2^m}{m}\right].$$

Solution:

Taking n = m, we first establish the reduction formula

$$I_{m,m} = \frac{1}{2m} + \frac{1}{2} I_{m-1,m-1} \qquad \text{...(1)}$$

$$\therefore I_{m,m} = \frac{1}{2m} + \frac{1}{2}\left[\frac{1}{2(m-1)} + \frac{1}{2} I_{m-2 \cdot m-2}\right]$$

$$\left[\because \text{from (1)}, I_{m-1,m-1} = \frac{1}{2(m-1)} + \frac{1}{2} I_{m-2,m-2}\right]$$

$$= \frac{1}{2m} + \frac{1}{2^2(m-1)} + \frac{1}{2^2} I_{m-2,m-2}$$

$$= \frac{1}{2m} + \frac{1}{2^2(m-1)} + \frac{1}{2^3(m-2)} + \frac{1}{2^3} I_{m-3,m-3}, \text{ and so on.}$$

Finally, $I_{m,m} = \dfrac{1}{2m} + \dfrac{1}{2^2(m-1)} + \dfrac{1}{2^3(m-2)}$

$$+ \ldots + \frac{1}{2^{m-2} \cdot 3} + \frac{1}{2^{m-1} \cdot 2} + \frac{1}{2^{m-1}} I_{1,1}$$

But $I_{1,1} = \int_0^{\pi/2} \cos x \sin x \, dx = \left[\frac{1}{2} \sin^2 x\right]_0^{\pi/2} = \frac{1}{2}$

$$\therefore \; I_{m,n} = \frac{1}{2^{m+1}}\left[2 + \frac{2^2}{2} + \frac{2^3}{2} + \ldots + \frac{2^m}{m}\right].$$

Example 13:

Prove that if n be a positive integer greater than unity, then $\int_0^{\pi/2} \cos^{n-2} x \sin nx \, dx = \frac{1}{n-1}$.

Solution:

Taking m = n − 2, we first establish the reduction formula

$$\int_0^{\pi/2} \cos^{n-2} x \sin nx \, dx$$

$$= \frac{1}{2n-2} + \frac{n-2}{2n-2}\int_0^{\pi/2} \cos^{n-3} x \sin (n-1)\, x \, dx$$

Applying this formula repeatedly, we have

$$\int_0^{\pi/2} \cos^{n-2} x \sin nx \, dx$$

$$= \frac{1}{2(n-1)} + \frac{n-2}{2(n-1)} \left\{\frac{1}{2n-4} + \frac{n-3}{2n-4}\int_0^{\pi/2} \cos^{n-4} x \sin (n-2)\, x \, dx\right\}$$

$$= \frac{1}{2(n-1)} + \frac{1}{2^2(n-1)} + \frac{n-3}{2^2(n-1)} \int_0^{\pi/2} \cos^{n-4} x \sin (n-2)\, x \, dx$$

$$= \frac{1}{2(n-1)} + \frac{1}{2^2(n-1)} + \frac{n-3}{2^2(n-1)} \left\{\frac{1}{2n-6} + \frac{n-4}{2n-6}\int_0^{\pi/2} \cos^{n-5} x \sin (n-3) x \, dx\right\}$$

and finally $= \frac{1}{(n-1)}\left[\frac{1}{2} + \frac{1}{2^2} + \frac{1}{2^3} + \ldots \frac{1}{2^{n-2}}\right] + \frac{1}{2^{n-2}(n-1)}\int_0^{\pi/2} (\cos x)^0 \sin 2x \, dx$

$$= \frac{1}{(n-1)} \cdot \frac{\frac{1}{2}\left\{1 - \left(\frac{1}{2}\right)^{n-2}\right\}}{1 - \frac{1}{2}} + \frac{1}{2^{n-2}(n-1)}\left[-\frac{1}{2}\cos 2x\right]_0^{\pi/2}$$

$$= \frac{1}{(n-1)}\left[1 - \frac{1}{2^{n-2}}\right] + \frac{1}{2^{n-2} \cdot (n-1)} \cdot 1$$

$$= \frac{1}{(n-1)}\left[1 - \frac{1}{2^{n-2}} + \frac{1}{2^{n-2}}\right] = \frac{1}{n-1}$$

Example 14:

Prove that if n be a positive integer,

$$\int_0^{\pi/2} \cos^n x \cos nx \, dx = \frac{\pi}{2^{n+1}}$$

Solution:

Taking m = n, we first establish the reduction formula

$$I_{m,n} = \frac{n}{n+n} I_{n-1,n-1} = \frac{1}{2} I_{n-1,n-1} \qquad ...(1)$$

Putting (n − 1) for n in (1),

we have $I_{n-1,n-1} = \frac{1}{2} I_{n-2,n-2}$.

$$\therefore \; I_{n,n} = \frac{1}{2} \cdot \frac{1}{2} I_{n-2,n-2} = \frac{1}{2^2} I_{n-2,n-2}$$

Thus by repeated application of (1), we get

$$I_{n,n} = \frac{1}{2^n} I_{n-n,n-n} = \frac{1}{2^n} I_{0,0}$$

But $I_{0,0} = \int_0^{\pi/2} \cos^0 x \cos 0x \, dx$

$$= \int_0^{\pi/2} 1 \cdot dx = [x]_0^{\pi/2} = \frac{\pi}{2}$$

$$\therefore \; I_{n,n} = \frac{1}{2^n} \cdot \frac{1}{2} \pi = \frac{\pi}{2^{n+1}}.$$

Example 15:

If $I_{(m,n)} = \int_0^{\pi/2} \cos^m x \cos nx \, dx$, *prove that*

$$I_{(m,n)} = \left\{\frac{m(m-1)}{m^2 - n^2}\right\} I_{(m-2,n)}$$

Solution:

We have $I_{(m,n)} = \int_0^{\pi/2} \cos^m x \cos nx \, dx$

$$= \left[\cos^m x \frac{\sin nx}{n}\right]_0^{\pi/2} - \int_0^{\pi/2} \frac{\sin nx}{n} m \cos^{m-1} x(-\sin x)\, dx$$

integrating by parts taking cos nx as the 2nd function

$$= 0 + \frac{m}{n}\int_0^{\pi/2} \cos^{m-1} x \sin nx \sin x \, dx$$

Again integrating by parts taking sin nx as the 2nd function, we have

$$I_{m,n} = \frac{m}{n}\left[(\cos^{m-1} x \sin x)\left(\frac{-\cos nx}{n}\right)\right]_0^{\pi/2}$$

$$-\frac{m}{n}\int_0^{\pi/2}\left(-\frac{\cos nx}{n}\right)[\cos^{m-1} x \cos x - (m-1)\cos^{m-2} x \sin^2 x]\, dx$$

$$= 0 + \frac{m}{n^2}\int_0^{\pi/2} \cos nx \,\{\cos^m x - (m-1)\cos^{m-2} x \cdot (1 - \cos^2 x)\}\, dx$$

$$= \frac{m}{n^2}\int_0^{\pi/2} \cos nx \,\{m \operatorname{cox}^m x - (m-1)\cos^{m-2} x\}\, dx$$

$$= \frac{m}{n^2}\{I_{m,n.} - (m-1) I_{m-2,n}\}$$

$$\therefore \left(1 - \frac{m^2}{n^2}\right) I_{m,n} = -\frac{m(m-1)}{n^2} I_{m-2,n}$$

or $$I_{m,n} = \frac{m(m-1)}{m^2 - n^2} I_{m-2,n}.$$

Example 16:

Find the reduction formula for the integral $\int \frac{\sin nx}{\sin x} dx$ *and show that* $\int_0^{\pi} \frac{\sin nx}{\sin x} dx = \pi$ *or 0, according as n is odd or even.*

Solution:

To find the required reduction formula, consider

$\sin nx - \sin(n-2)x = 2\cos(n-1)x \sin x$

or $\dfrac{\sin nx}{\sin x} - \dfrac{\sin(n-2)x}{\sin x} = 2\cos(n-1)x$, dividing both sides by sin x

or $\dfrac{\sin nx}{\sin x} = 2\cos(n-1)x + \dfrac{\sin(n-2)x}{\sin x}$

Integrating both the sides, we have

$$\int \frac{\sin nx}{\sin x}\,dx = \frac{2\sin(n-1)x}{(n-1)} + \int \frac{\sin(n-2)x}{\sin x}\,dx,$$

which is the required reduction formula.

Now let $I_n = \int_0^{\pi} \dfrac{\sin nx}{\sin x}\,dx$

$$= \left\{\frac{2\sin(n-1)x}{n-1}\right\}_0^{\pi} + \int_0^{\pi} \frac{\sin(n-2)x}{\sin x}\,dx = 0 + I_{n-2}$$

Hence $I_n = I_{n-2} = I_{n-4} = I_{n-6} = \ldots$

i.e., when n is even $I_n = I_2$

$$= \int_0^{\pi} \frac{\sin 2x}{\sin x}\,dx = 2\int_0^{\pi} \cos x\,dx = 2\,[\sin x]_0^{\pi} = 0$$

and when n is odd $I_n = I_1$

$$= \int_0^{\pi} \frac{\sin x}{\sin x}\,dx = \int_0^{\pi} dx = \pi.$$

Example 17:

Prove that $\displaystyle\int_0^{\pi} \left(\frac{\sin n\theta}{\sin\theta}\right)^2 d\theta = n\pi$

Solution:

Let $I_n = \displaystyle\int_0^{\pi} \left(\frac{\sin n\theta}{\sin\theta}\right)^2 d\theta$, then

$$I_{n-1} = \int_0^{\pi} \left(\frac{\sin(n-1)\theta}{\sin\theta}\right)^2 d\theta$$

$$\therefore\ I_n - I_{n-1} = \int_0^{\pi} \frac{\sin^2 n\theta - \sin^2 (n-1)\theta}{\sin^2 \theta}\, d\theta$$

$$= \int_0^{\pi} \frac{\sin(2n-1)\,\theta \cdot \sin\theta}{\sin^2\theta}\, d\theta = \int_0^{\pi} \frac{\sin(2n-1)\theta}{\sin\theta}\, d\theta \qquad \text{(Note)}$$

$= \pi,$

$$\left[\because \text{ from Ex. } 56 \int_0^{\pi} \frac{\sin n\theta}{\sin\theta}\, d\theta = \pi \text{ if n is odd, here } (2n-1) \text{ is odd}\right]$$

Hence $I_n = I_{n-1} + \pi$ ...(1)

$= I_{n-2} + 2\pi;$ $[\because \text{ from (1), } I_{n-1} = I_{n-2} + \pi]$

$= I_{n-3} + 3\pi$, and so on.

Thus $I_n = (n-1)\pi + I_1$

$$= (n-1)\pi + \int_0^{\pi} \left(\frac{\sin\theta}{\sin\theta}\right)^2 d\theta = (n-1)\pi + \pi = n\pi$$

Example 18:

If $S_n = \int_0^{\pi/2} \frac{\sin(2n-1)x}{\sin x}\, dx$

$V_n = \int_0^{\pi/2} \left(\frac{\sin nx}{\sin x}\right)^2 dx$, *(n is an integer), show that*

$S_{n+1} - S_n = 0,\ V_{n+1} - V_n = S_{n+1}$

Solution:

S_{n+1} and V_{n+1} will be obtained by writing $(n + 1)$ in place of n in S_n and V_n respectively. Thus

$$S_{n+1} = \int_0^{\pi/2} \frac{\sin(2n+1)\,x\, dx}{\sin x}$$

and $$V_{n+1} = \int_0^{\pi/2} \left\{\frac{\sin(n+1)x}{\sin x}\right\}^2 dx$$

$$\therefore\ S_{n+1} - S_n = \int_0^{\pi/2} \frac{[\sin(2n+1)\,x - \sin(2n-1)\,x]\, dx}{\sin x}$$

$$= \int_0^{\pi/2} \frac{\cos 2nx \text{ in } x\, dx}{\sin x}$$

$$= \int_0^{\pi/2} 2\cos 2nx\, dx = \left[\frac{2\sin 2nx}{2n}\right]_0^{\pi/2} = 0$$

($\because$ $\sin n\pi = 0$ when n is an integer and $\sin 0 = 0$)

$$\text{Also } V_{n+1} - V_n = \int_0^{\pi/2}\left[\left(\frac{\sin(n+1)x}{\sin x}\right)^2 - \left(\frac{\sin nx}{\sin x}\right)^2\right] dx$$

$$= \int_0^{\pi/2} \frac{[\sin^2 (n+1)x - \sin^2 (nx)]\, dx}{\sin^2 x}$$

$$= \int_0^{\pi/2} \frac{\{2\sin^2 (n+1)x - 2\sin^2 (nx)\}\, dx}{2\sin^2 x}$$

$$= \int_0^{\pi/2} \frac{\{1 - \cos 2(n+1)x - 1 + \cos 2nx\}\, dx}{2\sin^2 x}$$

$$= \int_0^{\pi/2} \frac{\{\cos 2nx - \cos 2(n+1)x\}\, dx}{2\sin^2 x}$$

$$= \int_0^{\pi/2} \frac{2\sin(2n+1)x \sin x\, dx}{2\sin^2 x} = \int_0^{\pi/2} \frac{\sin(2n+1)x}{\sin x}\, dx = S_{n+1}$$

Example 19:

Evaluate $\int_0^1 (\sin^{-1} x)^4\, dx$

Solution:

Put $\sin^{-1} x = t$

i.e., $x = \sin t$,

so that $dx = \cos t\, dt$.

When $x = 0$,

$t = \sin^{-1} 0 = 0$

and when $x = 1$,

$t = \sin^{-1} 1 = \frac{1}{2}\pi$

$\therefore$ the given integral $I = \int_0^{\pi/2} t^4 \cos t\, dt$

$= \left[t^4 \sin t\right]_0^{\pi/2} - \int_0^{\pi/2} 4t^2 \sin t\, dt$, integrating by parts taking $\cos t$ as the second function

$$= \frac{\pi^4}{16} - 4\int_0^{\pi/2} t^2 \sin t\, dt$$

$$= \frac{\pi^4}{16} - 4\left[\left\{t^3 (-\cos t)\right\}_0^{\pi/2} + \int_0^{\pi/2} 3t^2 \cos t\, dt\right]$$

again integrating by parts

$$= \frac{\pi^4}{16} - 4 \times 0 - 12\int_0^{\pi/2} t^2 \cos t\, dt$$

$$= \frac{\pi^4}{16} - 12\left[\left\{t^2 \tan t\right\}_0^{\pi/2} - \int_0^{\pi/2} 2t \sin t\, dt\right]$$

$$= \frac{\pi^4}{16} - 12\left(\frac{\pi^2}{4}\right) + 24\int_0^{\pi/2} t \sin t\, dt$$

$$= \frac{\pi^4}{16} - 3\pi^2 + 24\left[\left\{t(-\cos t)\right\}_0^{\pi/2} + \int_0^{\pi/2} \cos t\, dt\right]$$

$$= \frac{\pi^4}{16} - 3\pi^2 + 24 \times 0 + 24\left[\sin t\right]_0^{\pi/2}$$

$$= \frac{1}{16}\pi^4 - 3\pi^2 + 24.$$

Example 20:

Evaluate the following integrals:

(i) $\int_1^{\infty} \frac{x^4 + 1}{x^2 (x^2 + 1)^2}\, dx$

(ii) $\int_1^{\infty} \frac{x^2 + 3}{x^6 (x^2 + 1)^2}\, dx$

Solution:

(i) Let $I = \int_1^{\infty} \frac{x^4 + 1}{x^2 (x^2 + 1)^2}\, dx = \int_1^{\infty} \frac{(x^2 + 1)^2 - 2x^2}{x^2 (x^2 + 1)^2}\, dx$

$$= \int_1^{\infty} \frac{1}{x^2}\, dx - 2\int_1^{\infty} \frac{1}{(x^2 + 1)}\, dx$$

$$= \left[-\frac{1}{x}\right]_0^{\infty} - 2\int_0^{\infty} \frac{dx}{(x^2 + 1)^2} = 1 - 2\int_1^{\infty} \frac{dx}{(x^2 + 1)^2}$$

Now put $x = \tan t$,

so that $dx = \sec^2 t\, dt$.

When $x = 1,\ t = \frac{1}{4}\pi$

and $x = \infty,\ t = \frac{1}{2}\tau$.

$$\therefore\ I = 1 - 2\int_{\pi/4}^{\pi/2} \frac{\sec^2 t\, dt}{(1+\tan^2 t)^2} = 1 - 2\int_{\pi/4}^{\pi/2} \frac{\sec^2 t}{\sec^4 t}\, dt$$

$$= 1 - 2\int_{\pi/4}^{\pi/2} \cos^2 t\, dt = 1 - \int_{\pi/4}^{\pi/2} (1+\cos 2t)\, dt$$

$$= 1 - \left[t + \frac{1}{2}\sin 2t\right]_{\pi/4}^{\pi/2}$$

$$= 1 - \left[\left(\frac{1}{2}\pi + \frac{1}{2}\sin\pi\right) - \left(\frac{1}{4}\pi + \frac{1}{2}\sin\frac{1}{2}\pi\right)\right]$$

$$= 1 - \left[\frac{1}{2}\pi - \frac{1}{4}\pi - \frac{1}{2}\right] = 1 - \left[\frac{1}{4}\pi - \frac{1}{2}\right] = \frac{3}{2} - \frac{1}{4}\pi$$

(ii) The given integral $I = \int_1^\infty \frac{(x^2+1)+2}{x^6(x^2+1)}\, dx$

$$\int_1^\infty \frac{1}{x^6}\, dx + 2\int_1^\infty \frac{1}{x^6(x^2+1)}\, dx$$

$$= \left[-\frac{1}{5x^5}\right]_1^\infty + 2\int_1^\infty \frac{1}{x^6(x^2+1)}\, dx$$

$$= \frac{1}{5} + 2\int_1^\infty \frac{dx}{x^6(x^2+1)}.$$

Now put $x = \tan t$,

so that $dx = \sec^2 t\, dt$.

When $x = 1,\ t = \frac{\pi}{4}$

and when $x = \infty,\ t = \frac{\pi}{2}$.

$$\therefore\ I = \frac{1}{5} + 2\int_{\pi/4}^{\pi/2} \cot^6 t\, dt$$

Now we know that $\int \cot^n t\,dt = -\dfrac{\cot^{n-1} t}{n-1} - \int \cot^{n-2} t\,dt$

$$\therefore \int \cot^6 t\,dt = -\frac{\cot^5 t}{5} - \int \cot^4 t\,dt$$

$$= -\frac{\cot^5 t}{5} - \left[-\frac{\cot^3 t}{3} - \int \cot^2 t\,dt\right]$$

$$= -\frac{1}{5}\cot^5 t + \frac{1}{3}\cot^3 t + \int (\text{cosec}^2 t - 1)\,dt$$

$$= -\frac{1}{5}\cot^5 t + \frac{1}{3}\cot^3 t - \cot t - t$$

$$\therefore\ I = \frac{1}{5} + 2\left[-\frac{1}{5}\cot^5 t + \frac{1}{3}\cot^3 t - \cot t - t\right]_{\pi/4}^{\pi/2}$$

$$= \frac{1}{5} + 2\left[-\frac{1}{2}\pi - \left(-\frac{1}{2} + \frac{1}{3} - 1 - \frac{1}{4}\pi\right)\right]$$

$$= \frac{1}{5} + \frac{2}{5} - \frac{2}{3} + 2 + \left(-\pi + \frac{1}{2}\pi\right)$$

$$= \frac{29}{15} - \frac{1}{2}\pi = \frac{1}{30}(58 - 15\pi).$$

Example 21:

Establish a reduction formula for $\int \sin^n(2x)dx$.

Solution:

Let $I_n = \int \sin^n(2x)dx$ or $I_n = \int \sin^{n-1}(2x)\sin(2x)\,dx$.

Integrating by parts regarding sin 2x as the 2nd function, we have

$$I_n = \sin^{n-1}(2x)\left[-\frac{1}{2}\cos 2x\right]$$

$$-\int\left\{(n-1)\sin^{n-2}2x.\cos 2x.2\right\}.\left(-\frac{1}{2}\cos 2x\right)dx$$

$$= -\frac{1}{2}\sin^{n-1}2x.\cos 2x + (n-1)\int \sin^{n-2}2x.\cos^2 2x\,dx$$

$$= -\frac{1}{2}\sin^{n-1}2x.\cos 2x + (n-1)\int \sin^{n-2}2x.\left(1 - \sin^2 2x\right)dx$$

$$= -\frac{1}{2}\sin^{n-1}2x.\cos 2x + (n-1)\int \sin^{n-2}2x\,dx \;-(n-1)\int \sin^{n}2x\,dx$$

$$= -\frac{1}{2}\sin^{n-1}2x.\cos 2x + (n-1)I_{n-2} - (n-1)I_n.$$

Transposing the last term to the left, we have

$$nI_n = -\frac{1}{2}\sin^{n-1}2x.\cos 2x + (n-1)I_{n-2}$$

or $I_n = -\dfrac{\sin^{n-1}2x.\cos 2x}{2n} + \dfrac{n-1}{n}I_{n-2}$, is the reduction formula.

Example 22:

Prove that $\displaystyle\int_0^{\pi/2} \sin^{2m}x\,dx = \frac{(2m)!}{\{2^m m!\}^2}\cdot\frac{\pi}{2}.$

Solution:

Here 2m is even. Hence from 5.2 (Case II), we get

$$\int_0^{\pi/2} \sin^{2m}x\,dx = \frac{(2m-1)(2m-3)\ldots 3.1}{(2m)(2m-2)\ldots 4.2}\cdot\frac{\pi}{2} \qquad \text{(Willi's formula)}$$

$$= \frac{2m(2m-1)(2m-2)\ldots 3.2.1}{\{2m(2m-2)\ldots 4.2\}^2}\cdot\frac{\pi}{2}.$$

[Multiplying Nr. & Dr. by 2m (2m – 2) (2m – 4) ... 4 . 2]

$$= \frac{(2m)!}{\{2^m\cdot m(m-1)(m-2)\ldots 2.1\}^2}\cdot\frac{\pi}{2} = \frac{(2m)!}{\{2^m.m!\}^2}\cdot\frac{\pi}{2}.$$

Example 23:

Evaluate $\int \sin^6 x\,dx.$

Solution:

We know that

$$\int \sin^n x\,dx = -\frac{\sin^{n-1}x\cos x}{n} + \frac{(n-1)}{n}\int \sin^{n-2}x\,dx$$

[Establish the formula here]

Taking n = 6 and applying the above formula successively, we have

$$\int \sin^6 x\,dx = -\frac{\sin^{6-1}x\cos x}{6} + \frac{6-1}{6}\int \sin^{6-2}x\,dx$$

$$= -\frac{1}{6}\sin^5 x\cos x + \frac{5}{6}\int \sin^4 x\,dx$$

$$= -\frac{1}{6}\sin^5 x\cos x + \frac{5}{6}\left[-\frac{1}{4}\sin^3 x\cos x + \frac{3}{4}\int \sin^2 x\,dx\right]$$

$$= -\frac{1}{6}\sin^5 x\cos x - \frac{5}{24}\sin^3 x\cos x + \frac{5}{8}\int \sin^2 x\,dx$$

$$= -\frac{1}{6}\sin^5 x\cos x - \frac{5}{24}\sin^3 x\cos x + \frac{5}{8}\int \frac{1}{2}(1-\cos 2x)\,dx$$

$$= -\frac{1}{6}\sin^5 x\cos x - \frac{5}{24}\sin^3 x\cos x + \frac{5}{8}\left(\frac{1}{2}x - \frac{1}{2}\cdot\frac{1}{2}\sin 2x\right)$$

$$= -\frac{1}{6}\sin^5 x\cos x - \frac{5}{24}\sin^3 x\cos x - \frac{5}{16}\sin x\cos x + \frac{5}{16}x.$$

Example 24:

Evaluate $\int_0^{\pi/2} \sin^6 x\,dx$.

Solution:

Here n = 6 (even), we get

$$\int_0^{\pi/2} \sin^6 x\,dx = \frac{5}{6}\cdot\frac{3}{4}\cdot\frac{1}{2}\cdot\frac{\pi}{2} = \frac{5\pi}{32}.$$

Example 25:

Evaluate $\int_0^{\pi/2} \cos^9 x\,dx$.

Solution:

Here n = 9 (odd), we get

$$\int_0^{\pi/2} \cos^9 x\,dx = \frac{8.6.4.2}{9.7.5.3.1}\cdot 1 = \frac{128}{315}.$$

Example 26:

Evaluate $\int_0^{\pi/2} \sin^{10} x\,dx$ *or* $\int_0^{\pi/2} \cos^{10} x\,dx$.

Solution:

Here n = 10 (even) in both the cases.

$$\therefore \int_0^{\pi/2} \sin^{10} x\,dx = \int_0^{\pi/2} \cos^{10} x\,dx$$

$$=\frac{9.7.5.3.1}{10.8.6.4.2}\cdot\frac{\pi}{2}=\frac{63\pi}{512}.$$

Example 27:

Show that $\int_0^a \frac{x^4}{\sqrt{(a^2-x^2)}}dx=\frac{3a^4\pi}{16}.$

Solution:

Put $x = a \sin\theta$,

so that $dx = a \cos\theta \, d\theta$.

Also when $x = 0$,

$\sin\theta = 0$ *i.e.*, $\theta = 0$

and when $x = a$, $\sin\theta = 1$

i.e., $\theta=\frac{\pi}{2}$.

Then $\int_0^a \frac{x^4 dx}{\sqrt{(a^2-x^2)}}=\int_0^{\pi/2}\frac{a^4\sin^4\theta\, a\cos\theta\, d\theta}{\sqrt{(a^2-a^2\sin^2\theta)}}$

$$=a^4\int_0^{\pi/2}\sin^4\theta\, d\theta=a^4\cdot\frac{3.1}{4.2}\cdot\frac{\pi}{2}=\frac{3\pi a^4}{16}.$$

Example 28:

Evaluate $\int_0^{2a}\frac{x^{9/2}dx}{\sqrt{(2a-x)}}.$

Solution:

Put $x = 2a \sin^2\theta$,

so that $dx = 2a.2 \sin\theta \cos\theta \, d\theta$.

Also when $x = 0$,

$\sin^2\theta = 0$ *i.e.*, $\theta = 0$

and when $x = 2a$,

$\sin^2\theta = 1$ *i.e.*, $\theta=\frac{\pi}{2}$.

Then $\int_0^{2a}\frac{x^{9/2}dx}{\sqrt{(2a-x)}}=\int_0^{\pi/2}\frac{(2a\sin^2\theta)^{9/2}.4a\sin\theta\cos\theta\, d\theta}{\sqrt{(2a-2a\sin^2\theta)}}$

$$=\int_0^{\pi/2}\frac{(2a)^{9/2}.4a\sin^{10}\theta.\cos\theta\, d\theta}{(2a)^{1/2}.\cos\theta}=(2a)^4.4a\int_0^{\pi/2}\sin^{10}\theta\, d\theta$$

$$= 64a^5 \cdot \frac{9}{10} \cdot \frac{7}{8} \cdot \frac{5}{6} \cdot \frac{3}{4} \cdot \frac{1}{2} \cdot \frac{\pi}{2} = \frac{63a^5\pi}{8}.$$

Example 29:

Evaluate $\int_0^{\pi/4} \tan^5 \theta d\theta.$

Solution:

We have

$$\int \tan^n \theta d\theta = \frac{\tan^{n-1}\theta}{n-1} - \int \tan^{n-2}\theta d\theta. \qquad ...(1)$$

Putting n = 5 in (1), we have

$$\int \tan^5 \theta d\theta = \frac{1}{4}\tan^4\theta - \int \tan^3\theta d\theta$$

$$= \frac{1}{4}\tan^4\theta - \left[\frac{1}{2}\tan^2\theta - \int \tan\theta d\theta\right], \text{ putting } n = 3 \text{ in (1)}$$

$$= \frac{1}{4}\tan^4\theta - \frac{1}{2}\tan^2\theta - \log\cos\theta.$$

$$\therefore \int_0^{\pi/4} \tan^5\theta d\theta = \left[\frac{1}{4}\tan^4\theta - \frac{1}{2}\tan^2\theta - \log\cos\theta\right]_0^{\pi/4}$$

$$= \left[\frac{1}{4} - \frac{1}{2} - \log\cos\frac{1}{4}\pi\right] - [0 - \log\cos 0] = \left[-\frac{1}{4} - \log\left(\frac{1}{\sqrt{2}}\right)\right]$$

$$= \left[-\frac{1}{4} + \frac{1}{2}\log 2\right] = \frac{1}{2}\left[\log 2 - \frac{1}{2}\right].$$

Example 30:

Evaluate $\int_0^a x^5 \left(2a^2 - x^2\right)^{-3} dx.$

Solution:

Put $x = \sqrt{2}.a \sin\theta$

so that dx = $\sqrt{2}.a \cos\theta \, d\theta$

Also when x = 0,

sin θ = 0 or θ = 0

and when x = a,

sin θ = $\frac{1}{\sqrt{2}}$ or $\theta = \frac{\pi}{4}$.

Making these substitutions the given integral

$$= \int_0^{\pi/4} (\sqrt{2}.a \sin\theta)^5.(2a^2 - 2a^2\sin^2\theta)^{-3}.\sqrt{2}.a \cos\theta\, d\theta$$

$$= \int_0^{\pi/4} \left[\frac{2^{5/2} a^5 \sin^5\theta . 2^{1/2} a \cos\theta}{2^3 a^6 \cos^6\theta}\right] d\theta$$

$$= \int_0^{\pi/4} \tan^5\theta\, d\theta = \frac{1}{2}\left[\log 2 - \frac{1}{2}\right].$$

Example 31:

If $I_n = \int_0^{\pi/4} \tan^n x\, dx$, *show that*

$I_n + I_{n-2} = \frac{1}{n-1}$, *and deduce the value of* I_5.

Solution:

We know that

$$\int \tan^n x\,dx = \frac{\tan^{-1} x}{n-1} - \int \tan^{n-2} x\,dx \quad \text{(derive it here)}$$

$$\therefore\ I_n = \int_0^{\pi/4} \tan^n x\,dx = \left[\frac{\tan^{n-1} x}{n-1}\right]_0^{\pi/4} - \int_0^{\pi/4} \tan^{n-2} x\,dx$$

$$= \frac{1}{n-1} - I_{n-2}, \qquad \left[\because I_{n-2} = \int_0^{\pi/4} \tan^{n-2} x\,dx\right]$$

or $I_n + I_{n-2} = \frac{1}{n-1}$.

Putting n = 5 in the reduction formula $I_n = \frac{1}{n-1} - I_{n-2}$, we get

$$I_5 = \frac{1}{4} - I_3 = \frac{1}{4} - \left[\frac{1}{2} - I_1\right] = \frac{1}{4} - \frac{1}{2} + I_1$$

$$= -\frac{1}{4} + \int_0^{\pi/4} \tan x\,dx = -\frac{1}{4} + [\log \sec x]_0^{\pi/4}$$

$$= -\frac{1}{4} + \left[\log\sec\frac{\pi}{4} - \log\sec 0\right] = -\frac{1}{4} + [\log\sqrt{2} - \log 1]$$

$$= \frac{1}{2}\log 2 - \frac{1}{4} = \frac{1}{2}\left(\log 2 - \frac{1}{2}\right).$$

Example 32:

If $I_n = \int_0^{\pi/4} \tan^n x\, dx$, *prove that*

$n\,(I_{n+1} + I_{n-1}) = 1.$

Solution:

We have

$$I_{n+1} = \int_0^{\pi/4} \tan^{n+1} x\,dx = \left[\frac{\tan^n x}{n}\right]_0^{\pi/4} - \int_0^{\pi/4} \tan^{n-1} x\,dx$$

$$= \left[\frac{1}{n}\right] - I_{n-1}$$

or $\left[I_{n+1} + I_{n-1}\right] = \dfrac{1}{n}$

or $n\,(I_{n+1} + I_{n-1}) = 1.$

Similarly, $I_n + I_{n-2} = \dfrac{1}{(n-1)}$.

Example 33:

Evaluate $\int sec^3 x\,dx$.

Solution:

We have $\int \sec^3 x\,dx = \int \sec x.\sec^2 x\,dx$

$= \sec x \tan x - \int \sec x \tan x \tan x\,dx,$

(integrating by parts taking $\sec^2 x$ as the 2nd function)

$= \sec x \tan x - \int \sec x \tan^2 x\,dx$

$= \sec x \tan x - \int \sec x \left(\sec^2 x - 1\right)dx$

$= \sec x \tan x - \int \sec^3 x\,dx + \int \sec x\,dx.$

Transposing the term $-\int \sec^3 x\,dx$ to the left, we have

$2\int \sec^3 x\,dx = \sec x \tan x + \int \sec x\,dx$

or $\int \sec^3 x\,dx = \dfrac{1}{2}\sec x \tan x + \dfrac{1}{2}\log(\sec x + \tan x).$

Example 34:

Evaluate $\int_0^{\pi/4} sec^3 x\,dx$.

Solution:

This integral can be evaluated by proceeding in the same way as in part (a) of this exercise.

Otherwise, let $I = \int_0^{\pi/4} \sec^3 x\,dx = \int_0^{\pi/4} \sec x.\sec^2 x\,dx$

$$= \int_0^{\pi/4} \sqrt{\left(1+\tan^2 x\right)}.\sec^2 x\,dx$$

Now put $\tan x = t$,

so that $\sec^2 x\,dx = dt$.

When $x = 0, t = 0$

and when $x = \frac{1}{4}\pi, t = 1.$

$$\therefore\ I = \int_0^1 \sqrt{\left(1+t^2\right)}dt = \left[\frac{1}{2}t\sqrt{\left(1+t^2\right)} + \frac{1}{2}\log\left\{t+\sqrt{\left(1+t^2\right)}\right\}\right]_0^1$$

$$= \left[\frac{1}{2}\sqrt{2} + \frac{1}{2}\log\left(1+\sqrt{2}\right)\right] - \left[0+\frac{1}{2}\log 1\right]$$

$$= \frac{1}{2}\sqrt{2} + \frac{1}{2}\log\left(\sqrt{2}+1\right).$$

Example 35:

Evaluate $\int \sqrt{\left(\frac{x+a}{x}\right)}\,dx.$

Solution:

Put $x = a\tan^2\theta$, so that $dx = 2a\tan\theta\sec^2\theta\,d\theta$.

$$\text{Thus } \int \sqrt{\left(\frac{x+a}{x}\right)}dx = \int \sqrt{\left(\frac{a\sec^2\theta}{a\tan^2\theta}\right)} \cdot 2a\tan\theta\sec^2\theta\,d\theta$$

$$= 2a\int \sec^3\theta\,d\theta.$$

Example 36:

Evaluate $\int \frac{d\theta}{\sin^4 \frac{1}{2}\theta}.$

Solution:

We have $\int \frac{d\theta}{\sin^4 \frac{1}{2}\theta} = \int \operatorname{cosec}^4 \frac{\theta}{2}\,d\theta$

$$= 2\int \operatorname{cosec}^4 x\,dx, \text{ putting } \theta = 2x.$$

But $\int \operatorname{cosec}^n x\, dx = \dfrac{\operatorname{cosec}^{n-2} x \cot x}{n-1} + \dfrac{n-2}{n-1}\int \operatorname{cosec}^{n-2} x\, dx$

[Derive this formula here]

Putting n = 4, we get

$$\int \operatorname{cosec}^n x\, dx = -\frac{\operatorname{cosec}^2 x \cot x}{3} + \frac{2}{3}\int \operatorname{cosec}^2 x\, dx$$

$$= -\frac{1}{3}\operatorname{cosec}^2 x \cot x + \frac{2}{3}(-\cot x)$$

Hence the given integral

$$= 2\int \operatorname{cosec}^4 x\, dx = -\frac{2}{3}\operatorname{cosec}^2 x \cot x - \frac{4}{3}\cot x$$

$$= -\frac{2}{3}\operatorname{cosec}^2 \frac{1}{2}\theta \cot\frac{1}{2}\theta - \frac{4}{3}\cot\frac{1}{2}\theta. \quad [\because x = \theta/2]$$

Example 37:

Evaluate $\int (1+x^2)^{3/2}\, dx$.

Solution:

Put $x = \tan\theta$,

so that $dx = \sec^2\theta\, d\theta$.

Then $\int (1+x^2)^{3/2}\, dx = \int \sec^2\theta \sec^3\theta\, d\theta = \int \sec^5\theta\, d\theta$.

Now we shall form a reduction formula for $\int \sec^n\theta\, d\theta$. we get

$$\int \sec^n\theta\, d\theta = \frac{\sec^{n-2}\theta \tan\theta}{n-1} + \frac{n-2}{n-1}\int \sec^{n-2}\theta\, d\theta$$

$$\therefore \quad \int \sec^n\theta\, d\theta = \frac{1}{4}\sec^3\theta\tan\theta + \frac{3}{4}\int \sec^3\theta\, d\theta$$

$$= \frac{1}{4}\sec^3\theta\tan\theta + \frac{3}{4}\left[\frac{1}{2}\sec\theta\tan\theta + \frac{1}{2}\int \sec\theta\, d\theta\right]$$

$$= \frac{1}{4}\sec^3\theta\tan\theta + \frac{3}{8}\sec\theta\tan\theta + \frac{3}{8}\log(\sec\theta + \tan\theta)$$

$$= \frac{1}{4}[(1+x^2)^{3/2}\cdot x] + \frac{3}{8}x(1+x^2)^{1/2} + \frac{3}{8}\log\{x + \sqrt{(1+x^2)}\}.$$

Example 38:

Evaluate $\int_0^a (a^2 + x^2)^{5/2}\, dx$.

Solution:

Put $x = a \tan\theta$,

so that $dx = a \sec^2\theta\, d\theta$

$$\text{Then } I = \int_0^a (a^2 + x^2)^{5/2}\, dx$$

$$= \int_0^{\pi/4} a^4 \sec^5\theta\, a \sec^2\theta\, d\theta = a^6 \int_0^{\pi/4} \sec^7\theta\, d\theta$$

Now form a reduction formula for $\int \sec^n\theta\, d\theta$. By repeated application of this formula, we get

$$I = a^6\left[\left(\frac{1}{6}\sec^5\theta\tan\theta\right)_0^{\pi/4} + \frac{5}{6}\int_0^{\pi/4}\sec^5\theta\, d\theta\right]$$

$$= a^6\left[\frac{4\sqrt{2}}{6} + \frac{5}{6}\left(\frac{\sec^3\theta\tan\theta}{4}\right)_p^{\pi/4} + \frac{5}{6}\cdot\frac{3}{4}\int_0^{\pi/4}\sec^3\theta\, d\theta\right]$$

$$= a^6\left[\frac{2\sqrt{2}}{3} + \frac{5\sqrt{2}}{12} + \frac{5}{8}\cdot\frac{1}{2}(\sec\theta\tan\theta)_0^{\pi/4} + \frac{5}{8}\cdot\frac{1}{2}\int_0^{\pi/4}\sec\theta\, d\theta\right]$$

$$= a^6\left[\frac{2\sqrt{2}}{3} + \frac{5\sqrt{2}}{12} + \frac{5\sqrt{2}}{16} + \frac{5}{16}\left\{\log\tan\left(\frac{1}{4}\pi + \frac{1}{2}\theta\right)\right\}_0^{\pi/4}\right]$$

$$= a^6\left[\frac{2\sqrt{2}}{3} + \frac{5\sqrt{2}}{12} + \frac{5\sqrt{12}}{16} + \frac{5}{16}\log\tan\left(\frac{3}{8}\pi\right)\right]$$

$$= a^6\left[\frac{67\sqrt{2}}{48} + \frac{5}{16}\log\tan\left(\frac{3}{8}\pi\right)\right]$$

$$= \frac{a^6}{48}\left[67\sqrt{2} + 15\log\tan\left(\frac{3}{8}\pi\right)\right].$$

Example 39:

Evaluate $\int_0^2 (4 + x^2)^{5/2}\, dx$.

Solution:

Do your self.

Example 40:

Evaluate $\int_0^{\pi/4} \sin^2 \theta \cos^4 \theta d\theta$

Solution:

We have the reduction formula

$$\int \sin^m x \cos^n x \, dx = \frac{\sin^{m+1} x \cdot \cos^{n-1} x}{m+n} + \frac{n-1}{m+1} \int \sin^m x \cos^{n-2} x \, dx$$

[Derive it here]

Here m = 2 and n = 4; hence we have

$$\int_0^{\pi/4} \sin^2 \theta \cos^4 \theta \, d\theta = \left[\frac{\sin^3 \theta \cos^3 \theta}{6}\right]_0^{\pi/4} + \frac{3}{6}\int_0^{\pi/4} \sin^2 \theta \cos^2 \theta \, d\theta$$

$$= \frac{1}{48} + \frac{1}{2}\left[\left(\frac{\sin^3 \theta \cos\theta}{4}\right)_0^{\pi/4} + \frac{1}{4}\int_0^{\pi} \sin^2 \theta \, d\theta\right]$$

[Putting m = 2 and n = 2 in the above reduction formula]

$$= \frac{1}{48} + \frac{1}{32} + \frac{1}{8} \cdot \frac{1}{2} \int_0^{\pi/4} (1 - \cos 2\theta) \, d\theta$$

$$= \frac{1}{48} + \frac{1}{32} + \frac{1}{16}\left[\theta - \frac{\sin 2\theta}{2}\right]_0^{\pi/4}$$

$$= \frac{1}{48} + \frac{1}{32} + \frac{1}{16}\left[\frac{\pi}{4} - \frac{1}{2}\right] = \frac{1}{48} + \frac{\pi}{64}.$$

Example 41:

Evaluate $\int \frac{\sin^2 x}{\cos^3 x} dx.$

Solution:

The given integral $I = \int \sin^2 x \, (\cos x)^{-3} \, dx.$

Here m = 2 and n = – 3 (negative), therefore we will apply the reduction formula in which n increases *i.e.*,

$$\int \sin^m x \cos^n x \, dx = \frac{\sin^{m+1} x \cos^{n+1} x}{n+1} + \frac{m+n+2}{n+1} \int \sin^m x \cos^{n+2} x \, dx$$

$$\therefore \int \sin^2 x(\cos x)^{-3}\,dx = -\frac{\sin^3 x \cos^{-2} x}{-2} + \left(-\frac{1}{2}\right)\int \sin^2 x(\cos x)^{-1}\,dx$$

$$= \frac{1}{2}\frac{\sin^3 x}{\cos^2 x} - \frac{1}{2}\int \frac{1-\cos^2 x}{\cos x}\,dx$$

$$= \frac{1}{2}\frac{\sin^3 x}{\cos^2 x} - \frac{1}{2}\int (\sec x - \cos x)\,dx$$

$$= \frac{1}{2}\frac{\sin^3 x}{\cos^2 x} - \frac{1}{2}\log(\sec x + \tan x) + \frac{1}{2}\sin x$$

$$= \frac{1}{2}\sin\left(\frac{\sin^3 x}{\cos^2 x} + 1\right) - \frac{1}{2}\log(\sec x + \tan x)$$

$$= \frac{1}{2}[\sec x \tan x - \log(\sec x + \tan x)].$$

Example 42:

Evaluate $\int_0^{\pi/2} \sin^4 x \cos^2 x\,dx.$

Solution:

We know that

$$\int_0^{\pi/2} \sin^m x \cos^n x\,dx = \frac{\Gamma\left(\frac{m+1}{2}\right)\Gamma\left(\frac{n+1}{2}\right)}{2\Gamma\left(\frac{m+n+1}{2}\right)}$$

$$\therefore \text{ The given integral} = \frac{\Gamma\left(\frac{4+1}{2}\right)\cdot\Gamma\left(\frac{2+1}{2}\right)}{2\Gamma\left(\frac{4+2+1}{2}\right)} = \frac{\Gamma\frac{5}{2}\,\Gamma\frac{3}{2}}{2\Gamma 4}$$

$$= \frac{\frac{3}{2}\cdot\frac{1}{2}\sqrt{\pi}\cdot\frac{1}{2}\sqrt{\pi}}{2\cdot 3\cdot 2\cdot 1} = \frac{\pi}{32} \qquad \left[\because \Gamma(n+1) = n\Gamma n \text{ and } \Gamma\frac{1}{2} = \sqrt{\pi}\right].$$

Alternative Solution : Using Walli's formula, the given integral

$$= \frac{3\cdot 1\cdot 1}{6\cdot 4\cdot 2}\cdot\frac{\pi}{2} = \frac{\pi}{32}.$$

Example 43:

Evaluate $\int_0^{\pi/2} \sin^2 x \cos^3 x\, dx$

Solution:

Here m = 2, n = 3;

using the Gamma function, we have the given integral

$$= \frac{\Gamma\left(\frac{2+1}{2}\right)\cdot\Gamma\left(\frac{3+1}{2}\right)}{2\Gamma\left(\frac{2+3+2}{2}\right)} = \frac{\Gamma\frac{3}{2}\cdot\Gamma 2}{2\Gamma\frac{7}{2}} = \frac{\frac{1}{2}\sqrt{\pi}\cdot 1}{2\cdot\frac{5}{2}\cdot\frac{3}{2}\cdot\frac{1}{2}\sqrt{\pi}} = \frac{2}{15}$$

Otherwise : By Walli's formula, the given integral $= \frac{1\cdot 2}{5\cdot 3\cdot 1}\times 1 = \frac{2}{15}.$

Example 44:

Evaluate $\int_0^{\pi/2} \sin^4 x \cos^6 x\, dx$

Solution:

Here m = 4, n = 6;

using the Gamma function, we have the given integral

$$= \frac{\Gamma\left(\frac{4+1}{2}\right)\cdot\Gamma\left(\frac{6+1}{2}\right)}{2\Gamma\left(\frac{4+6+2}{2}\right)} = \frac{\Gamma\frac{5}{2}\cdot\Gamma\frac{7}{2}}{2\Gamma 6}$$

$$= \frac{\frac{3}{2}\cdot\frac{1}{2}\sqrt{\pi}+\frac{5}{2}\cdot\frac{3}{2}\cdot\frac{1}{2}\sqrt{\pi}}{2\cdot 5\cdot 4\cdot 3\cdot 2\cdot 1} = \frac{3\pi}{512}$$

Otherwise. By Walli's formula, the given integral

$$= \frac{3\cdot 1\cdot 5\cdot 3\cdot 1}{10\cdot 8\cdot 6\cdot 4\cdot 2}\cdot\frac{\pi}{2} = \frac{3\pi}{512}.$$

Example 45:

Evalueate $\int_0^{\pi/2} \sin^5 x \cos^8 x\, dx$

Solution:

Here m = 5, n = 8;

using the Gamma function, we have the given integral

$$= \frac{\Gamma\left(\frac{5+1}{2}\right)\Gamma\left(\frac{8+1}{2}\right)}{2\Gamma\left(\frac{5+8+2}{2}\right)} = \frac{\Gamma(3)\cdot\Gamma\left(\frac{9}{2}\right)}{2\Gamma\left(\frac{15}{2}\right)}$$

$$= \frac{2\cdot 1\cdot \Gamma\frac{9}{2}}{2\cdot\frac{13}{2}\cdot\frac{11}{2}\cdot\frac{9}{2}\Gamma\frac{9}{2}} = \frac{2\cdot 2\cdot 2\cdot 2}{2\cdot 13\cdot 11\cdot 9} = \frac{8}{1287}.$$

Example 46:

Evaluate $\int_0^{\pi/2} \sin^{12} x \cos^{18} x\, dx$

Solution:

Using the Gamma function, we have the given integral

$$= \frac{\Gamma\left(\frac{12+1}{2}\right)\cdot\Gamma\left(\frac{18+1}{2}\right)}{2\Gamma\left(\frac{12+18+2}{2}\right)} = \frac{\Gamma\left(\frac{13}{2}\right)\Gamma\left(\frac{19}{2}\right)}{2\Gamma(16)}$$

Integrals reducible to the form $\int_0^{\pi/2} \sin^m x \cos^n x\, dx$.

Example 47:

Evaluate $\int_0^{\pi/2} \sin^6 \theta d\theta$

Solution:

We have $\int_0^{\pi/2} \sin^6 \theta\, d\theta = \int_0^{\pi/2} \sin^6 \theta \cos^0 \theta\, d\theta$ (Note)

Now m = 6, n = 0;

using the Gamma function, we have the given integral

$$= \frac{\Gamma\left(\frac{6+1}{2}\right)\cdot\Gamma\left(\frac{0+1}{2}\right)}{2\Gamma\left(\frac{6+0+2}{2}\right)} = \frac{\Gamma\frac{7}{2}\cdot\Gamma\frac{1}{2}}{2\Gamma 4}$$

$$= \frac{\frac{5}{2}\cdot\frac{3}{2}\cdot\frac{1}{2}\sqrt{\pi}\cdot\sqrt{\pi}}{2\cdot(3\cdot 2\cdot 1)} = \frac{5\pi}{32}$$

Otherwise, by Walli's formula, the given integral

$$= \frac{5 \cdot 3 \cdot 1}{6 \cdot 4 \cdot 2} \cdot \frac{\pi}{2} = \frac{5\pi}{32}$$

Example 48:

Evaluate $\int_0^{\pi/8} \cos^3 4x \, dx$

Solution:

To bring the given integral into the form of Gamma function, put 4x = θ, so that 4dx = dθ. Also for limits,

$$\theta = 0 \text{ at } x = 0$$

and $\theta = \frac{1}{2}\pi$ at $x = \frac{1}{8}\pi$

$\therefore$ the given integral $= \frac{1}{4}\int_0^{\pi/2} \cos^3 \theta \, d\theta = \frac{1}{4}\int_0^{\pi/2} \sin^0 \theta \cdot \cos^3 \theta \, d\theta$

$$= \frac{1}{4} \cdot \frac{\Gamma\frac{1}{2} \cdot \Gamma 2}{2\Gamma\frac{5}{2}} = \frac{\Gamma\frac{1}{2} \cdot 1}{4 \cdot 2 \cdot \frac{3}{2} \cdot \frac{1}{2} \Gamma\frac{1}{2}} = \frac{1}{6}.$$

Example 49:

Evaluate $\int_0^{\pi/2} \sin^6 \frac{x}{2} \cos^8 \frac{x}{2} \, dx$

Solution:

Putting x/2 = θ,

so that dx = 2dθ, the given integral

$$= 2\int_0^{\pi/2} \sin^6 \theta \cos^8 \theta \, d\theta$$

$$= 2 \cdot \frac{5 \cdot 3 \cdot 1 \cdot 7 \cdot 5 \cdot 3 \cdot 1}{14 \cdot 12 \cdot 10 \cdot 8 \cdot 6 \cdot 4 \cdot 2} \cdot \frac{\pi}{2} = \frac{5\pi}{2048}.$$

Example 50:

Evaluate $\int_0^{\pi/6} \sin^2 6\theta \cos^5 3\theta \, d\theta$

Solution:

To bring the given integral into the form of Gamma function, put 3θ = x, so that 3dθ = dx. Also for limits, x = 0 at θ = 0 and x = π/2 at θ = π/6.

$\therefore$ the given integral $= \frac{1}{3}\int_0^{\pi/2} \sin^2 2x \cos^5 x\, dx$

$$= \frac{1}{3}\int_0^{\pi/2} (2\sin x \cos x)^2 \cos^5 x\, dx = \frac{4}{3}\int_0^{\pi/2} \sin^2 x \cos^7 x\, dx$$

$$= \frac{4}{3}\frac{\Gamma\frac{3}{2}\cdot\Gamma 4}{2\Gamma\frac{11}{2}} = \frac{4}{3}\cdot\frac{\Gamma\frac{3}{2}\cdot 3\cdot 2\cdot 1}{2\cdot\frac{9}{2}\cdot\frac{7}{2}\cdot\frac{5}{2}\cdot\frac{3}{2}\cdot\Gamma\frac{3}{2}} = \frac{64}{945}.$$

Example 51:

Show that $\int_0^{\pi/6} \cos^4 3\phi \sin^3 6\phi\, d\phi = \frac{1}{15}$

Solution:

Do your self.

Example 52:

Evaluate $\int_0^{\pi/4} (\cos 2\theta)^{3/2} \cos\theta\, d\theta$

Solution:

The given integral $= \int_0^{\pi/4} (1 - 2\sin^2\theta)^{3/2} \cos\theta\, d\theta.$ (Note)

Now put $\sqrt{2}\sin\theta = \sin x$,

so that $\sqrt{2}\cos\theta\, d\theta = \cos x\, dx$.

Also when $\theta = 0$,

$\sin x = \sqrt{2}\sin 0 = 0$ giving $x = 0$

and when $\theta = \frac{1}{4}\pi$;

$\sin x = \sqrt{2}\sin\left(\frac{\pi}{4}\right) = 1$ giving $x = \frac{1}{2}\pi$.

Hence the given integral $= \int_0^{\pi/2} (1 - \sin^2 x)^{3/2} \cdot \frac{1}{\sqrt{2}}\cos x\, dx$

$$= \frac{1}{\sqrt{2}}\int_0^{\pi/2} \cos^3 x \cdot \cos x\, dx = \frac{1}{\sqrt{2}}\int_0^{\pi/2} \cos^4 x\, dx$$

$$= \frac{1}{\sqrt{2}}\int_0^{\pi/2} \sin^0 x \cdot \cos^4 x\, dx$$

$$=\frac{1}{\sqrt{2}}\cdot=\frac{\Gamma\left(\frac{0+1}{2}\right)\cdot\Gamma\left(\frac{4+1}{2}\right)}{2\Gamma\left(\frac{0+4+2}{2}\right)}=\frac{1}{\sqrt{2}}\frac{\Gamma\frac{5}{2}\cdot\Gamma\frac{1}{2}}{2\Gamma 3}$$

$$=\frac{1}{\sqrt{2}}\cdot\frac{\frac{3}{2}\cdot\frac{1}{2}\sqrt{\pi}\sqrt{\pi}}{2\cdot 2\cdot 1}=\frac{3\pi}{16\sqrt{2}}.$$

Example 53:

Evaluate $\int_0^1 x^4(1-x^2)^{5/2}\,dx$.

Solution:

Here we put $x = \sin\theta$,
so that $dx = \cos\theta\, d\theta$.
And now the new limits are $\theta = 0$ to $\theta = \pi/2$.

Thus the given integral $= \int_0^{\pi/2}\sin^4\theta\,(1-\sin^2\theta)^{5/2}\cos\theta\, d\theta$

$$=\int_0^{\pi/2}\sin^4\theta\cdot\cos^5\theta\cos\theta\, d\theta=\int_0^{\pi/2}\sin^4\theta\cos^6\theta\, d\theta$$

$$=\frac{3\cdot 1\cdot 5\cdot 3\cdot 1}{10\cdot 8\cdot 6\cdot 4\cdot 2}\cdot\frac{\pi}{2}$$ [By Walli's formula]

$$=\frac{3\pi}{512}.$$

Example 54:

Evaluate $\int_0^{\pi/2}\cos^5 x\sin 3x\, dx$.

Solution:

The given integral $I = \int_0^{\pi/2}\cos^5 x\,(3\sin x - 4\sin^3 x)\, dx$

$$=3\int_0^{\pi/2}\cos^5 x\sin x\, dx-4\int_0^{\pi/2}\cos^5 x\sin^3 x\, dx$$

$$=3\cdot\frac{4\cdot 2}{6\cdot 4\cdot 2}-4\cdot\frac{4\cdot 2\cdot 2}{8\cdot 6\cdot 4\cdot 2}$$

$$=\frac{1}{2}-\frac{1}{6}=\frac{1}{3}.$$

Example 55:

Evaluate $\int_0^{\pi/2} \sin^3 x \cos^4 x \cos 2x \, dx \, m$

Solution:

$$I = \int_0^{\pi/2} \sin^3 x \cos^4 x (\cos^2 x - \sin^2 x) \, dx$$

$$= \int_0^{\pi/2} \sin^3 x \cos^6 x \, dx - \int_0^{\pi/2} \sin^5 x \cos^4 x \, dx$$

$$= \frac{2 \cdot 5 \cdot 3 \cdot 1}{9 \cdot 7 \cdot 5 \cdot 3 \cdot 1} - \frac{4 \cdot 2 \cdot 3 \cdot 1}{9 \cdot 7 \cdot 5 \cdot 3 \cdot 1} = \frac{2}{63} - \frac{8}{315} = \frac{10-8}{315} = \frac{2}{315}.$$

Example 56:

Show that $\int_0^1 x^2 (1-x^2)^{3/2} \, dx = \frac{\pi}{32}$

Solution:

We have (*i.e.* x = sin θ),

The given integral = $\int_0^{\pi/2} \sin^2 \theta \cos^3 \theta \cdot \cos\theta \, d\theta$

$$= \int_0^{\pi/2} \sin^2 \theta \cos^4 \theta \, d\theta$$

$$= \frac{1 \cdot 3 \cdot 1}{6 \cdot 4 \cdot 2} \cdot \frac{\pi}{2}$$ [By Walli's formula]

$$= \frac{\pi}{32}.$$

Example 57:

Evaluate $\int_0^1 x^4 (1-x^2)^{3/2} \, dx$

Solution:

Putting x = sin θ etc., we get the given integral =

$$\int_0^{\pi/2} \sin^4 \theta \cdot \cos^3 \theta \cdot \cos \theta \, d\theta = \int_0^{\pi/2} \sin^4 \theta \cos^4 \theta \, d\theta$$

$$= \frac{3 \cdot 1 \cdot 3 \cdot 1}{8 \cdot 6 \cdot 4 \cdot 2} \cdot \frac{\pi}{2},$$ [By Walli's formula]

$$= \frac{3\pi}{256}.$$

Example 58:

Show that $\int_0^a x^4(a^2 - x^2)^{1/2}\,dx = \dfrac{\pi a^6}{32}$.

Solution:

Put x = a sin θ

so that dx = a cos θ dθ.

Also when x = 0, θ = 0

and when x = a, $\theta = \dfrac{1}{2}\pi$

$$\therefore \text{ the given integral } = \int_0^{\pi/2} a^4 \sin^4\theta \cdot a\cos\theta \cdot a\cos\theta\, d\theta$$

$$= a^6\int_0^{\pi/2} \sin^4\theta\cos^2\theta\, d\theta = a^6\,\frac{\Gamma\frac{5}{2}\cdot\Gamma\frac{3}{2}}{2\Gamma 4}$$

$$= a^6\,\frac{\frac{3}{2}\cdot\frac{1}{2}\sqrt{\pi}\,\frac{1}{2}\sqrt{\pi}}{2\cdot 3\cdot 2\cdot 1} = \frac{\pi a^6}{32}.$$

Example 59:

Evaluate $\int_0^1 x^6(1 - x^2)^{1/2}\,dx$

Solution:

Putting x = sin θ etc., we get,

The given integral

$$= \int_0^{\pi/2} \sin^6\theta\sqrt{(1-\sin^2\theta)\cos\theta}\, d\theta$$

$$= \int_0^{\pi/2} \sin^6\theta\cos^2\theta\, d\theta$$

$$= \frac{5\cdot 3\cdot 1\cdot 1}{8\cdot 6\cdot 4\cdot 2}\cdot\frac{\pi}{2} \qquad \text{[By Walli's formula]}$$

$$= \frac{5\pi}{256}.$$

Example 60:

Show that $\int_0^a x^2(a^2 - x^2)^{3/2}\,dx = \dfrac{\pi a^6}{32}$

Solution:

Putting x = a sin θ we get the given integral

$$= a^6 \int_0^{\pi/2} \sin^2\theta \cos^4\theta \, d\theta = \frac{\pi a^6}{32}.$$

Example 61:

Evaluate $\int_0^1 x^m (1-x)^n \, dx$.

Solution:

Here we put $x = \sin^2\theta$,

so that $dx = 2 \sin\theta \cos\theta \, d\theta$.

Also when $x = 0$, $\theta = 0$

and when $x = 1$, $\theta = \frac{1}{2}\pi$

$\therefore$ the given integral $= \int_0^{\pi/2} \sin^{2m}\theta \cdot \cos^{2n}\theta \cdot 2\sin\theta\cos\theta \, d\theta$

$$= 2\int_0^{\pi/2} \sin^{2m+1}\theta \cos^{2n+1}\theta \, d\theta$$

$$= 2 \cdot \frac{\Gamma\left\{\frac{1}{2}(2m+2)\right\}\Gamma\left\{\frac{1}{2}(2n+2)\right\}}{2\Gamma\left\{\frac{1}{2}(2m+2n+4)\right\}} = \frac{\Gamma(m+1)\,\Gamma(n+1)}{\Gamma(m+n+2)}.$$

Example 62:

Show that $\int_0^1 x^{3/2} (1-x)^{3/2} \, dx = \frac{3\pi}{128}$

Solution:

Put $x = \sin^2\theta$,

so that $dx = 2 \sin\theta \cos\theta \, d\theta$.

Also when $x = 0$, $\theta = 0$

and when $x = 1$, $\theta = \frac{1}{2}\pi$

$\therefore$ the given integral

$$= \int_0^{\pi/2} \sin^3\theta \cos^3\theta \cdot 2\sin\theta\cos\theta \, d\theta$$

$$= \int_0^{\pi/2} \sin^4\theta \cos^4\theta \, d\theta$$

$$= 2\,\frac{3\cdot 1\cdot 3\cdot 1}{8\cdot 6\cdot 4\cdot 2}\cdot\frac{\pi}{2}$$ [by Walli's formula]

$$= \frac{3\pi}{128}.$$

Example 63:

Evaluate $\int_0^1 x^{3/2} \sqrt{(1-x)}\, dx$

Solution:

The given integral reduces to

$$2\int_0^{\pi/2} \sin^4\theta \cos^2\theta \, d\theta = 2 \cdot \frac{3\cdot 1\cdot 1}{6\cdot 4\cdot 2}\cdot\frac{\pi}{2} = \frac{\pi}{16}$$

Example 64:

Prove that $\int_0^1 \frac{dx}{\sqrt{(1-x^n)}} = \frac{\sqrt{\pi}\,\Gamma(1/n)}{n\Gamma\left\{\frac{1}{2}+(1/n)\right\}}$

Solution:

Put $x^n = \sin^2\theta$ *i.e.*, $x = (\sin\theta)^{2/n}$

so that $dx = \left(\frac{2}{n}\right)(\sin\theta)^{\left(\frac{2}{n}\right)-1} \cdot \cos\theta\, d\theta$

Also $\theta = 0$ when $x = 0$

and $\theta = \frac{1}{2}\pi$ when $x = 1$

$\therefore$ the given integral

$$= \frac{2}{n}\int_0^{\pi/2} \frac{(\sin\theta)^{\left(\frac{2}{n}\right)-1} \cdot \cos\theta\, d\theta}{\cos\theta}$$

$$= \frac{2}{n}\int_0^{\pi/2} (\sin\theta)^{\left(\frac{2}{n}\right)-1} d\theta = \frac{2}{n}\int_0^{\pi/2} (\sin\theta)^{\left(\frac{2}{n}\right)-1} \cos^0\theta\, d\theta$$

$$= \frac{2\Gamma\frac{1}{2}\Gamma\left(\frac{1}{n}\right)}{n\, 2\Gamma\left\{\frac{1}{2}+\left(\frac{1}{n}\right)\right\}} = \frac{\sqrt{\pi}\cdot\Gamma(1/n)}{n\Gamma\left\{\frac{1}{2}+\left(\frac{1}{n}\right)\right\}}.$$

Example 65:

Evaluate $\int_0^{2a} x^m \sqrt{(2ax-x^2)}\, dx$, *m being a positive integer.*

Solution:

We have $\int_0^{2a} x^m \sqrt{(2ax - x^2)}\, dx$

$= \int_0^{2a} x^m \cdot x^{1/2} \sqrt{(2a - x)}\, dx$

Now put $x = 2a \sin^2 \theta$,

so that $dx = 4a \sin\theta \cos\theta\, d\theta$.

Also when $x = 0, \theta = 0$

and when $x = 2a, \theta = \frac{1}{2}\pi$.

$\therefore$ The given integral,

$$= \int_0^{\pi/2} (2a \sin^2\theta)^m (2a \sin^2\theta)^{1/2}, \sqrt{(2a \cos^2\theta)} \cdot 4a \sin\theta \cos\theta\, d\theta$$

$$= 2^{m+3} a^{m+2} \int_0^{\pi/2} \sin^{2m+2}\theta \cos^2\theta\, d\theta$$

$$= 2^{m+3} a^{m+2} \frac{\Gamma\left(\frac{2m+3}{2}\right)\Gamma\frac{3}{2}}{\Gamma\left(\frac{2m+2+2+2}{2}\right)}$$

$$= 2^{m+3} a^{m+2} \frac{\frac{2m+1}{2} \cdot \frac{2m-1}{2} \cdots \frac{3}{2} \cdot \frac{1}{2} \cdot \sqrt{x} \cdot \frac{1}{2} \sqrt{\pi}}{(m+2)(m+1)m\,(m-1)\ldots 2 \cdot 1}$$

$$= a^{m+2} \frac{(2m+1)(2m-1)\ldots 3 \cdot 1}{(m+2)(m+1)\, m\, (m-1)\ldots 2 \cdot 1}.$$

Example 66:

Evaluate $\int_0^{2a} x^5 \sqrt{(2ax - x^2)}\, dx$

Solution:

The given integral

$$= 2^8 a^7 \int_0^{\pi/2} \sin^{12}\theta \cos^2\theta\, d\theta$$

$$= 2^8 a^7 \frac{11.9.7.5.3.1.1}{14.12.10.8.6.4.2} \cdot \frac{\pi}{2} = \frac{33}{16}\pi a^7$$

Example 67:

Evaluate $\int_0^a x^3 \sqrt{(2ax - x^2)^{3/2}}\, dx$

Solution:

We have

$$\int_0^a x^3 (2ax - x^2)^{3/2}\, dx = \int_0^a x^3 \cdot x^{3/2} (2a - x)^{3/2}\, dx$$

$$= \int_0^a x^{9/2} (2a - x)^{3/2}\, dx$$

Now put $x = 2a \sin^2 \theta$,

so that $dx = 4a \sin\theta \cos\theta\, d\theta$.

Also when $x = 0$,

$2a \sin^2 \theta = 0$

or $\sin\theta = 0$

i.e. $\theta = 0$

and when $x = a$,

$2a \sin^2\theta = a$

or sins $\theta = \dfrac{1}{\sqrt{2}}$ *i.e.*, $\theta = \dfrac{\pi}{4}$.

$\therefore$ the given integral

$$= \int_0^{\pi/4} (2a \sin^2\theta)^{9/2} (2a)^{3/2} (\cos^2\theta)^{3/2} \cdot 4a \sin\theta \cos\theta\, d\theta$$

$$= \int_0^{\pi/4} (2a)^{9/2} \sin^9\theta\, (2a)^{3/2} \cdot \cos^3\theta \cdot 4a \sin\theta \cos\theta\, d\theta$$

$$= (2a)^6 \cdot 4a \int_0^{\pi/4} \sin^{10}\theta \cdot \cos^4\theta\, d\theta.$$

This is not Gamma function as the limits are from 0 to $\pi/2$. We shall reduce it to the form of Gamma function by suitable trigonometrical adjustment. Thus the given integral

$$= a^7 \cdot 2^8 \int_0^{\pi/4} (\sin^2\theta)^3 (\sin^4\theta \cos^4\theta)\, d\theta \qquad \text{(Note)}$$

$$= 2a^7 \int_0^{\pi/4} (2\sin^2\theta)^3 \cdot (2\sin\theta\cos\theta)^4\, d\theta$$

$$= 2a^7 \int_0^{\pi/4} (1 - \cos 2\theta)^3 \sin^4 2\theta\, d\theta$$

Now put $2\theta = \alpha$,

so that $2d\theta = d\alpha$ and the new limits are $\alpha = 0$ to $\alpha = \pi/2$.

$\therefore$ the given integral

$$= 2a^7 \int_0^{\pi/2} (1 - \cos\alpha)^3 \sin^4\alpha \cdot \frac{1}{2}\, d\alpha$$

$$= a^7 \int_0^{\pi/2} (1 - \cos\alpha)^3 \sin^4\alpha\, d\alpha$$

$$= a^7 \int_0^{\pi/2} (1 - 3\cos\alpha + 3\cos^2\alpha - \cos^3\alpha) \sin^4\alpha\, d\alpha$$

$$= a^7 \int_0^{\pi/2} (\sin^4\alpha - 3\sin^4\alpha\cos\alpha + 3\sin^4\alpha\cos^2\alpha - \cos^3\alpha\sin^4\alpha)\, d\alpha$$

$$= a^7 \left[\frac{3 \cdot 1}{4 \cdot 2} \cdot \frac{\pi}{2} - 3 \cdot \frac{3 \cdot 1 \cdot 1}{5 \cdot 3 \cdot 1} \cdot 1 + 3 \cdot \frac{3 \cdot 1 \cdot 1}{6 \cdot 4 \cdot 2} \cdot \frac{\pi}{2} - \frac{2 \cdot 3 \cdot 1}{7 \cdot 5 \cdot 3 \cdot 1} \cdot 1 \right]$$

by Walli's formula

$$= a^7 \left(\frac{3\pi}{16} + \frac{3\pi}{32} - \frac{3}{5} - \frac{2}{35} \right) = a^7 \left(\frac{9\pi}{32} - \frac{23}{35} \right).$$

Example 68:

Evaluate $\displaystyle\int_0^a \frac{x^4}{(x^2 + a^2)^4}\, dx$

Solution:

Put $x = a \tan\theta$,

so that $dx = a \sec^2\theta\, d\theta$.

When $x = 0$, $\theta = 0$

and when $x = a$, $\tan\theta = 1$

i.e., $\theta = \frac{1}{4}\pi$

$\therefore$ the given integral $I = \displaystyle\int_0^{\pi/4} \frac{a^4 \tan^4\theta}{(a^2 \tan^2\theta + a^2)^4}\, a \sec^2\theta\, d\theta$

$$= \frac{1}{a^3} \int_0^{\pi/4} \frac{\tan^4\theta}{\sec^8\theta} \sec^2\theta\, d\theta = \frac{1}{a^3} \int_0^{\pi/4} \sin^4\theta \cos^2\theta\, d\theta$$

Now we get

$$I = \frac{1}{a^3} \cdot \frac{1}{16} \left[\frac{\pi}{4} - \frac{1}{3} \right].$$

Example 69:

Evaluate $\displaystyle\int_0^a x^2 \sqrt{(ax - x^2)}\, dx$

Solution:

We have $\int_0^a x^2 \sqrt{(ax - x^2)}\, dx = \int_0^a x^{5/2} \sqrt{(a - x)}\, dx$

Now put $x = a \sin^2 \theta$,

so that $dx = 2a \sin\theta \cos\theta\, d\theta$,

and the new limits are $\theta = 0$ to $\theta = \pi/2$.

Thus the given integral

$$= \int_0^{\pi/2} (a \sin^2 \theta)^{5/2} (a \cos^2 \theta)^{1/2} \cdot 2a \sin\theta \cos\theta\, d\theta$$

$$= 2a^4 \int_0^{\pi/2} \sin^6 \theta \cos^2 \theta\, d\theta = 2a^4 \frac{\Gamma\frac{7}{2} \cdot \Gamma\frac{3}{2}}{2\Gamma 5}$$

$$= 2a^4 \left[\frac{\frac{5}{2} \cdot \frac{3}{2} \cdot \frac{1}{2} \sqrt{\pi} \cdot \frac{1}{2} \sqrt{\pi}}{2 \cdot 4 \cdot 3 \cdot 2 \cdot 1} \right] = \frac{5\pi a^4}{128}.$$

Example 70:

Evaluate the following integrals:

(i) $\int_0^a x^2 \sqrt{\left(\frac{a - x}{a + x}\right)}\, dx$

(ii) $\int_0^a x \sqrt{\left(\frac{a^2 - x^2}{a^2 + x^2}\right)}\, dx$

Solution:

(i) Let $I = \int_0^a x^2 \sqrt{\left(\frac{a - x}{a + x}\right)}\, dx = \int_0^a \frac{x^2 (a - x)}{\sqrt{(a^2 - x^2)}}\, dx$

Now put $x = a \sin\theta$

so that $dx = a \cos\theta\, d\theta$.

Also $\theta = 0$ when $x = 0$

and $\theta = \theta = \frac{1}{2}\pi$ when $x = a$.

$$\therefore\ I = \int_0^{\pi/2} \frac{a^2 \sin^2 \theta\, (a - a \sin\theta)}{a \cos\theta}\, a \cos\theta\, d\theta$$

$$= a^3 \int_0^{\pi/2} (\sin^2\theta - \sin^3\theta)\, d\theta$$

$$= a^3 \left[\frac{1}{2}\cdot\frac{\pi}{2} - \frac{2}{3\cdot 1}\cdot 1\right]. \qquad \text{by Walli's formula}$$

$$= a^3\left(\frac{1}{4}\pi - \frac{2}{3}\right).$$

(ii) Let $I = \int_0^a x\sqrt{\left(\frac{a^2 - x^2}{a^2 + x^2}\right)}\, dx = \int_0^a \frac{x(a^2 - x^2)}{\sqrt{(a^4 - x^4)}}\, dx$

Now put $\qquad x^2 = a^2 \sin\theta$

so that $\qquad 2x\, dx = a^2 \cos\theta\, d\theta.$

Also $\theta = 0$ when $x = 0$

and $\theta = \frac{1}{2}\pi$ when $x = a$.

$$\therefore\ I = \int_0^{\pi/2} \frac{(a^2 - a^2\sin\theta)}{a^2\cos\theta}\cdot\frac{a^2}{2}\cos\theta\, d\theta$$

$$= \frac{a^2}{2}\int_0^{\pi/2}(1 - \sin\theta)\, d\theta = \frac{a^2}{2}[\theta + \cos\theta]_0^{\pi/2}$$

$$= \frac{a^2}{2}\left[\left(\frac{\pi}{2} + 0\right) - (0 + 1)\right] = \frac{1}{2}a^2\left(\frac{1}{2}\pi - 1\right) = \frac{1}{4}a^2(\pi - 2)$$

Example 71:

Show that $\int_0^\infty \frac{x^4\, dx}{(1 + x^2)^4} = \frac{\pi}{32}$

Solution:

Put $x = a\tan\theta$,

so that $dx = a\sec^2\theta\, d\theta$.

Also when $x = 0$, $\theta = 0$

and when $x = \infty$, $\theta = \pi/2$.

Hence the given integeral

$$= \int_0^{\pi/2} \frac{\tan^4\theta\sec^2\theta\, d\theta}{(1 + \tan^2\theta)^4} = \int_0^{\pi/2} \frac{\tan^4\theta\sec^2\theta\, d\theta}{\sec^8\theta}$$

$$= \int_0^{\pi/2} \sin^4\theta\cos^2\theta\, d\theta$$

$$= \frac{\Gamma\frac{5}{2}.\Gamma\frac{3}{2}}{2\Gamma 4} = \frac{\frac{3}{2}.\frac{1}{2}\sqrt{\pi}.\frac{1}{2}\sqrt{\pi}}{2.3.2.1} = \frac{\pi}{32}$$

Example 72:

Evaluate $\int_0^\infty \frac{x^4\,dx}{(a^2 + x^2)^4}$

Solution:

Put $x = a \tan \theta$,

so that $dx = a \sec^2 \theta \, d\theta$ and the new limits are $\theta = 0$ to $\theta = \pi/2$.

$\therefore$ the given integral

$$= \int_0^{\pi/2} \frac{a^4 \tan^4 \theta \cdot a \sec^2 \theta \, d\theta}{(a^2 + a^2 \tan^2 \theta)^4}$$

$$= \frac{1}{a^3}\int_0^{\pi/2} \frac{\tan^4 \theta \, d\theta}{\sec^6 \theta} = \frac{1}{a^3}\int_0^{\pi/2} \sin^4 \theta \cos^2 \theta \, d\theta$$

$$= \frac{1}{a^3}\frac{\Gamma\frac{5}{2}\cdot\Gamma\frac{3}{2}}{2\Gamma 4} = \frac{1}{a^3}\frac{\frac{3}{2}\cdot\frac{1}{2}\sqrt{\pi}\cdot\frac{1}{2}\sqrt{\pi}}{2\cdot 3\cdot 2\cdot 1} = \frac{\pi}{32a^3}$$

Example 73:

If m, n are positive integers, then prove that

$$\int_0^1 x^{m-1}(1-x)^{n-1}\,dx = \int_0^1 x^{n-1}(1-x)^{m-1}\,dx$$

$$= \frac{1\cdot 2\cdot 3\cdot \ldots (m-1)}{n(n+1)\ldots(n+m-1)}$$

Solution:

Put $\quad x = \sin^2 \theta$,

so that $\quad dx = 2 \sin \theta \cos \theta \, d\theta$.

Also when $\quad x = 0$,

$\theta = 0$

and when $\quad x = 1$,

$\theta = \pi/2$.

$$\therefore \int_0^1 x^{m-1}(1-x)^{n-1}\,dx$$

$$= \int_0^{\pi/2} \sin^{2m-2}\theta\,(1-\sin^2\theta)^{n-1} \cdot 2\sin\theta\cos\theta\, d\theta$$

$$= 2\int_0^{\pi/2} \sin^{2m-1}\theta \cos^{2n-1}\theta\, d\theta$$

$$= 2\,\frac{\Gamma\left\{\dfrac{2m-1+1}{2}\right\} \cdot \Gamma\left\{\dfrac{2n-1+1}{2}\right\}}{\Gamma\left\{\dfrac{2m-1+2n-1+2}{2}\right\}} = \frac{\Gamma m \cdot \Gamma n}{\Gamma(m+n)}$$

Thus $\displaystyle\int_0^1 x^{m-1}(1-x)^{n-1}\,dx = \frac{\Gamma(m)\cdot\Gamma(n)}{\Gamma(m+n)}$ ...(1)

Interchanging m and n in (1), we get

$$\int_0^1 x^{n-1}(1-x)^{m-1}\,dx = \frac{\Gamma n \cdot \Gamma m}{\Gamma(n+m)} \qquad ...(2)$$

From (1) and (2), we have

$$\int_0^1 x^{m-1}(1-x)^{n-1}\,dx = \int_0^1 x^{n-1}(1-x)^{m-1}\,dx$$

$$= \frac{\Gamma m \cdot \Gamma n}{\Gamma(m+n)} = \frac{(m-1)!\,(n-1)!}{(m+n-1)!}, \qquad [\because \Gamma r = (r-1)!]$$

$$= \frac{(m-1)!\,(n-1)!}{(n+m-1)(n+m-2)\ldots n\cdot(n-1)(n-2)\ldots 2\cdot 1} \qquad \text{(Note)}$$

$$= \frac{(m-1)!\,(n-1)!}{(n+m-1)(n+m-2)\ldots n\cdot(n-1)!}$$

$$= \frac{(m-1)(m-2)\ldots 3\cdot 2\cdot 1}{(n+m-1)(n+m-2)\ldots(n+1)\,n}.$$

Example 74:

Evaluate $\displaystyle\int_0^a x^2(2ax-x^2)^{5/2}\,dx$

Solution:

Putting $x = 2a\sin^2\theta$,

we get the given integral

$$= \int_0^{\pi/4} (2a\sin^2\theta)^{9/2}(2a)^{5/2}(\cos^2\theta)^{5/2}\, 4a\sin\theta\cos\theta\, d\theta$$

$$= 2^9 a^8 \int_0^{\pi/4} \sin^{10}\theta \cos^6\theta \, d\theta$$

$$= 2a^8 \int_0^{\pi/4} (2\sin^2\theta)^2 \, (2\sin\theta\cos\theta)^6 \, d\theta \qquad \text{(Note)}$$

$$= 2a^8 \int_0^{\pi/4} (1-\cos 2\theta)^2 \sin^6 2\theta \, d\theta$$

$$= a^8 \int_0^{\pi/2} (1-\cos\alpha)^2 \cdot \sin^6\alpha \, d\alpha, \text{ putting } 2\theta = \alpha$$

$$= a^8 \int_0^{\pi/2} (\sin^6\alpha - 2\cos\alpha \cdot \sin^6\alpha + \cos^2\alpha \sin^6\alpha) \, d\alpha$$

$$= a^8 \left[\frac{\Gamma\frac{7}{2}\cdot\Gamma\frac{1}{2}}{2\Gamma 4} - 2\frac{\Gamma 1\cdot\Gamma\frac{7}{2}}{2\Gamma\frac{9}{2}} + \frac{\Gamma\frac{3}{2}\cdot\Gamma\frac{7}{2}}{2\Gamma 5} \right] = a^8 \left[\frac{45\pi}{256} - \frac{2}{7} \right].$$

Example 75:

Evaluate $\int_0^{\pi/4} \sin^4 x \cos^2 x \, dx$

Solution:

The given integral $I = \int_0^{\pi/4} (\sin^2 x \cos^2 x) \sin^2 x \, dx$

$$= \int_0^{\pi/4} \frac{1}{4}(4\sin^2 x \cos^2 x) \cdot \frac{1}{2}(2\sin^2 x) \, dx$$

$$= \frac{1}{8}\int_0^{\pi/4} \sin^2 2x \, (1-\cos 2x) \, dx$$

Put $2x = t$,

so that $2dx = dt$.

Also when $x = 0$, $t = 0$

and when $x = \pi/4$, $t = \pi/2$.

$$\therefore \; I = \frac{1}{8}\int_0^{\pi/2} \sin^2 t \, (1-\cos t) \cdot \frac{1}{2} \, dt$$

$$= \frac{1}{16}\left[\int_0^{\pi/2} \sin^2 t \, dt - \int_0^{\pi/2} \sin^2 t \cos t \, dt \right]$$

$$= \frac{1}{16}\left[\frac{1}{2}\cdot\frac{\pi}{2} - \frac{1\cdot 1}{3\cdot 1} \right] = \frac{1}{16}\left[\frac{\pi}{4} - \frac{1}{3} \right].$$

Example 76:

Evaluate $\int_a^b (x-a)^m (b-x)^n \, dx$

Solution:

Here we put $x = b \sin^2\theta + a\cos^2\theta$,

so that $dx = 2(b-a)\sin\theta\cos\theta\, d\theta$.

Also when $x = a$, $(b-a)\sin^2\theta = 0$

i.e., $\theta = 0$

and when $x = b$, $(a-b)\cos^2\theta = 0$

i.e., $\theta = \pi/2$

Also $(x-a) = (b-a)\sin^2\theta$

and $(b-x) = (b-a)\cos^2\theta\, d\theta$

Thus the given integral

$$= \int_0^{\pi/2} (b-a)^m \sin^{2m}\theta \cdot (b-a)^n \cos^{2n}\theta \cdot 2(b-a)\sin\theta\cos\theta\, d\theta$$

$$= 2\int_0^{\pi/2} (b-a)^{m+n+1} \sin^{2m+1}\theta \cos^{2n+1}\theta\, d\theta$$

$$= (b-a)^{m+n+1}\left[\frac{\Gamma(m+1)\Gamma(n+1)}{\Gamma(m+n+2)}\right], \text{ applying Gamma function.}$$

Example 77:

Evaluate $\int_0^{\pi/2} x^3 \sin 3x \, dx$.

Solution:

Integrating by parts taking sin 3x as the 2nd function, we have the given integral

$$= \left[x^3 \cdot \frac{-\cos 3x}{3}\right]_0^{\pi/2} + \int_0^{\pi/2} 3x^2 \cdot \frac{\cos 3x}{3}\, dx$$

$$= 0 + \int_0^{\pi/2} x^2 \cos 3x\, dx$$

$$= \left[x^2 \frac{\sin 3x}{3}\right]_0^{\pi/2} - \int_0^{\pi/2} 2x \cdot \frac{\sin 3x}{3}\, dx$$

again integrating by parts regarding sin 3x as the 2nd function

$$= \frac{\pi^2}{4}\cdot\left(\frac{-1}{3}\right) - \frac{2}{3}\int_0^{\pi/2} x \sin 3x\, dx$$

$$= -\frac{\pi^2}{12} - \frac{2}{3}\left[\left(-x\,\frac{\cos 3x}{3}\right)_0^{\pi/2} + \int_0^{\pi/2}\frac{1}{3}\cos 3x\,dx\right]$$

$$= -\frac{\pi^2}{12} - \frac{2}{9}\int_0^{\pi/2}\cos 3x\,dx = -\frac{\pi^2}{12} - \frac{2}{9}\left[\frac{\sin 3x}{3}\right]_0^{\pi/2}$$

$$= -\frac{\pi^2}{12} + \frac{2}{27}.$$

Example 78:

Evaluate $\int_0^{\pi} x \sin^2 x \cos x \, dx$.

Solution:

Let $I = \int_0^{\pi} x \sin^2 x \cos x\,dx$. Intergrating by parts taking ($\sin^2 x \cos x$) as the 2nd function, we have

$$I = \left[\frac{x \sin^3 x}{3}\right]_0^{\pi} - \int_0^{\pi} 1\cdot\frac{\sin^3 x}{3}\,dx = -\frac{1}{3}\int_0^{\pi}\sin^3 x\,dx$$

$$= -\frac{1}{3}\int_0^{\pi}(1-\cos^2 x)\sin x\,dx = -\frac{1}{3}\left[-\cos x + \frac{1}{3}\cos^3 x\right]_0^{\pi}$$

$$= -\frac{1}{3}\left[\left(1-\frac{1}{3}\right)-\left(-1+\frac{1}{3}\right)\right] = -\frac{4}{9}.$$

Example 79:

Evaluate $\int_0^1 x^6 \sin^{-1} x\,dx$

Solution:

Put $\sin^{-1} x = t$

or $x = \sin t$,

so that $dx = \cos t\,dt$

$$\therefore\; I = \int_0^{\pi/2} t \sin^6 t \cos t\,dt$$

Integrating by parts taking $\sin^6 t \cos t$ as the second function and t as the first function, we have

$$I = \left[t\cdot\frac{\sin^7 t}{7}\right]_0^{\pi/2} - \int_0^{\pi/2} 1\cdot\frac{\sin^7 t}{7}\,dt$$

$$= \frac{\pi}{14} - \frac{1}{7} \cdot \frac{6 \cdot 4 \cdot 2}{7 \cdot 5 \cdot 3 \cdot 1} = \frac{\pi}{14} - \frac{16}{245}.$$

Example 80:

Evaluate $\int_0^a \sqrt{(a^2 - x^2)} \left\{ \cos^{-1}\left(\frac{x}{a}\right) \right\}^2 dx.$

Solution:

Here put $x = a \cos \theta,$

so that $dx = -a \sin \theta \, d\theta.$

Also when $x = 0,$

$\theta = \pi/2,$

and when $x = a,$

$\theta = 0,$

$\therefore$ the given integral

$$= -\int_{\pi/2}^0 (a \sin \theta)(\theta^2) \cdot a \sin \theta \, d\theta$$

$$= -a^2 \int_{\pi/2}^0 \theta^2 \sin^2 \theta \, d\theta = -a^2 \int_{\pi/2}^0 \theta^2 \cdot \frac{1}{2} \cdot 2 \sin^2 \theta \, d\theta$$

$$= -\frac{a^2}{2} \int_{\pi/2}^0 \theta^2 (1 - \cos 2\theta) \, d\theta$$

$$= \frac{a^2}{2} \int_{\pi/2}^0 \theta^2 \cos 2\theta \, d\theta - \frac{a^2}{2} \int_{\pi/2}^0 \theta^2 d\theta$$

$$= \frac{a^2}{2} \int_{\pi/2}^0 \theta^2 \cos 2\theta \, d\theta - \frac{a^2}{2} \left\{ \frac{1}{3} \theta^3 \right\}_{\pi/2}^0$$

$$= \frac{\pi^3 a^2}{48} + \frac{a^2}{2} \int_{\pi/2}^0 \theta^2 \cos 2\theta \, d\theta$$

Now to evaluate $\int \theta^2 \cos 2\theta \, d\theta$, applying integration by parts taking $\cos 2\theta$ as the 2nd function, we have

$$\int \theta^2 \cos 2\theta \, d\theta = \theta^2 \cdot \frac{1}{2} \sin 2\theta - \int 2\theta \cdot \frac{1}{2} \sin 2\theta \, d\theta$$

$$= \frac{1}{2} \theta^2 \sin 2\theta - \int \theta \sin 2\theta \, d\theta$$

$$= \frac{1}{2}\theta^2 \sin 2\theta - \left[\theta \cdot \left(-\frac{1}{2}\cos 2\theta\right) - \int 1 \cdot \left(-\frac{1}{2}\cos 2\theta\right) d\theta\right]$$

$$= \frac{1}{2}\theta^2 \sin 2\theta + \frac{1}{2}\theta \cos 2\theta - \frac{1}{4}\sin 2\theta$$

$$\therefore \int_{\pi/2}^{0} \theta^2 \cos 2\theta \, d\theta = \left[\frac{\theta^2 \sin 2\theta}{2} + \frac{\theta \cos 2\theta}{2} - \frac{\sin 2\theta}{4}\right]_{\pi/2}^{0} = \frac{\pi}{4}$$

$\therefore$ the required integral

$$= \frac{\pi^3 a^2}{48} + \frac{1}{2}a^2 \cdot \frac{\pi}{4} = \frac{\pi a^2}{8}\left(1 + \frac{1}{6}\pi^2\right).$$

Example 81:

If $u_n = \int_0^{\pi/2} x^n \sin x \, dx$ *and* $n > 1$,

show that $u_n + n(n-1)u_{n-2} = n\left(\frac{1}{2}\pi\right)^{n-1}$

Hence evaluate $\int_0^{\pi/2} x^5 \sin x \, dx$

Solution:

We have $u_n = \int_0^{\pi/2} x^n \sin x \, dx$

$$= \left[x^n \cdot (-\cos x)\right]_0^{\pi/2} - \int_0^{\pi/2} n \cdot x^{n-1} \cdot (-\cos x) \, dx$$

[Intergrating by parts taking sin x as the 2nd function]

$$= 0 + n\int_0^{\pi/2} x^{n-1} \cos x \, dx$$

$$= n\left[\left\{x^{n-1} \cdot \sin x\right\}_0^{\pi/2} - \int_0^{\pi/2} (n-1) \cdot x^{n-2} \cdot \sin x \, dx\right]$$

again integrating by parts

$$= n \cdot \left(\frac{1}{2}\pi\right)^{n-1} - n(n-1)\int_0^{\pi/2} x^{n-2} \sin x \, dx$$

Thus $u_n = n\left(\frac{1}{2}\pi\right)^{n-1} - n(n-1)u_{n} - 2$ ...(1)

$$\therefore\ u_n + n(n-1)u_{n-2} = n\left(\frac{1}{2}\pi\right)^{n-1}$$

Now to evaluate $\int_0^{\pi/2} x^5 \sin x\, dx$,

put n = 5 in (1).

Then $\quad u_5 = 5\left(\frac{1}{2}\pi\right)^{5-1} - 5(5-1)u_3$

$$= 5\cdot\left(\frac{1}{2}\pi\right)^4 - 20\left[3\left(\frac{1}{2}\pi\right)^{3-1} - 3(3-1)\,u_1\right],$$

putting n = 3 in (1)

$$= \frac{5}{16}\pi^4 - 15\pi^2 + 120u_1$$

Now $u_1 = \int_0^{\pi/2} x \sin x\, dx = \left[x\cdot(-\cos x)\right]_0^{\pi/2} + \int_0^{\pi/2} \cos x\, dx$

$$= \left[0 + \sin x\right]_0^{\pi/2} = \left[\sin\frac{\pi}{2} - \sin 0\right] = 1$$

Hence $u_5 = \int_0^{\pi/2} x^5 \sin x\, dx = \dfrac{5\pi^4}{16} - 15\pi^2 + 120.$

Example 82:

If $u_n = \int_0^{\pi/2} x^n \sin mx\, dx$,

prove that

$$u_n = \frac{n\pi^{n-1}}{m^2\cdot 2^{n-1}} - \frac{n(n-1)}{m^2}\,u_{n-2}.$$

if m is of the form 4r + 1.

Solution:

We have $\int_0^{\pi/2} x^n \sin mx\, dx$

$$= \left[x^n\cdot\left(-\frac{\cos mx}{m}\right)\right]_0^{\pi/2} - \int_0^{\pi/2} nx^{n-1}\cdot\left(-\frac{\cos mx}{m}\right)dx$$

$$= 0 + \frac{n}{m}\int_0^{\pi/2} x^{n-1} \cos mx \, dx$$

[$\because$ if m is of the form 4r + 1, then

$$\cos\{(4r+1)\pi/2\} = \cos\left(2r\pi + \frac{1}{2}\pi\right) = \cos\frac{1}{2}\pi = 0]$$

$$= \frac{n}{m}\left[\left\{x^{n-1} \cdot \frac{\sin mx}{m}\right\}_0^{\pi/2} - \int_0^{\pi/2} (n-1)x^{n-2} \cdot \left(\frac{\sin mx}{m}\right) dx\right],$$

again integrating by parts taking cos mx as 2nd function

$$= \frac{n}{m}\left[\left\{\left(\frac{\pi}{2}\right)^{n-1} \cdot \frac{1}{m}\right\} - \frac{(n-1)}{m}\int_0^{\pi/2} x^{n-2} \sin mx \, dx\right],$$

$$\left[\because \sin\frac{m\pi}{2} = \sin(4r+1)\frac{\pi}{2} = 1\right]$$

$$= \frac{n\pi^{n-1}}{m^2\, 2^{n-1}} - \frac{n(n-1)}{m^2} u_{n-2}.$$

3

REDUCTION FORMULAE CONTINUED

(For Irrational Algebraic and Transcendental Functions)

3.1. REDUCTION FORMULAE FOR $\int x^m (a + bx^n)^p \, dx$

$\int x^m (a + bx^n)^p \, dx$ can be connected with any one of the following six integrals:

(i) $\int x^{m-n} (a + bx^n)^p \, dx,$

(ii) $\int x^m (a + bx^n)^{p-1} \, dx,$

(iii) $\int x^{m+n} (a + bx^n)^p \, dx,$

(iv) $\int x^m (a + bx^n)^{p+1} \, dx$

(v) $\int x^{m-n} (a + bx^n)^{p+1} \, dx,$

(vi) $\int x^{m+n} (a + bx^n)^{p-1} \, dx.$

Rule for Connection

In order to connect the given integral with any one of the six integrals we use the following rule:

Let $P = x^{\lambda+1} (a + bx^n)^{\mu+1}$, where λ and μ are the smaller of the indices of x and $(a + bx^n)$ in the two expressions whose integrals we want to connect.

Find (dP/dx) and rearrange this as a linear combination of the expressions whose integrals are to be connected.

Finally integrate both sides and transpose suitably to get the required reduction formula.

(i) To connect $\int x^m (a + bx^n)^p \, dx$ with $\int x^{m-n} (a + bx^n)^p \, dx$

Here $\lambda = m - n$ and $\mu = p$ (choosing smaller indices for λ and μ)
Let us take $P = x^{\lambda+1} (a + bx^n)^{\mu+1} = x^{m-n+1}(a + bx^n)^{p+1}$.

$$\therefore \frac{dP}{dx} = (m-n+1)x^{m-n}(a+bx^n)^{p+1}$$

$$+ x^{m-n+1}(p+1)(a+bx^n)^p bnx^{n-1}$$

$= (m - n + 1)x^{m-n}(a + bx^n)^p(a + bx^n) + (p + 1)bnx^m(a + bx^n)^p$

$= a(m - n + 1)x^{m-n}(a + bx^n)^p + b(m - n + 1)x^m(a + bx^n)^p$

$+ (p + 1)bn\, x^m(a + bx^n)^p$

$= a(m - n + 1)x^{m-n}(a + bxn)p$

$+ bx^m(a + bx^n)^p(m - n + 1 + pn + n)$

i.e. $\left(\frac{dP}{dx}\right) = a(m-n+1)x^{m-n}(a+bx^n)^p$

$+ b(m+pn+1)x^m(a+bx^n)^p$

Thus $\left(\frac{dP}{dx}\right)$ is expressed as a linear combination of the two expressions whose integrals are to be connected.

Integrating both the sides, we have

$P = a(m-n+1)\int x^{m-n}(a+bx^n)^p\,dx$

$+ b(m+np+1)\int x^m(a+bx^n)^p\,dx$

Now putting the value of P and transposing suitably, we get

$b(pn+m+1)\int x^m(a+bx^n)^p\,dx$

$= x^{m-n+1}(a+bx^n)^{p+1} - a(m-n+1)\int x^{m-n}(a+bx^n)^p\,dx$

or $\int x^m(a+bx^n)^p\,dx$

$$= \frac{x^{m-n+1}(a+bx^n)^{p+1}}{b(pn+m+1)} - \frac{a(m-n+1)}{b(pn+m+1)}\int x^{m-n}(a+bx^n)^p\,dx$$

(ii) To connect $\int x^m(a+bx^n)^p\,dx$ with $\int x^m(a+bx^n)^{p-1}\,dx$

Here $\lambda = m$ and $\mu = p - 1$; (choosing smaller indices for λ and μ).
Now let $P = x^{\lambda+1}(a + bx^n)^{\mu+1} = x^{m+1}(a + bx^n)^p$.

$$\therefore \frac{dP}{dx} = (m+1)x^m(a+bx^n)^p + x^{m+1}\cdot p(a+bx^n)^{p-1}bn\,x^{n-1}$$

$= (m + 1)x^m(a + bx^n)^p + pn \times x^m(a + bx^n)^{p-1}bx^n$ (Note)

$= (m + 1)x^m(a + bx^n)^p + pn \times x^m(a + bx^n)^{p-1}(a + bx^n - a)$

$= (pn + m + 1)x^m(a + bx^n)^p - a\,pn\,x^m(a + bx^n)^{p-1}$.

Thus $\left(\frac{dP}{dx}\right)$ has been expressed as a linear combination of the two expressions whose integrals are to be connected.

Integrating both the sidses, we have

$$p = (pn + m + 1)\int x^m (a + bx^n)^p dx - apn\int x^m (a + bx^n)^{p=1} dx$$

Now putting the value of P, dividing by (pn + m + 1) and transposing suitably, we get

$$\int x^m (a + bx^n)^p dx = \frac{x^{m+1}(a + bx^n)^p}{np + m + 1}$$

$$+ \frac{anp}{np + m + 1}\int x^m (a + bx^n)^{p-1} dx$$

(iii) To connect $\int x^m (a + bx^n)^p dx$ with $\int x^{m+n} (a + bx^n)^p dx$

Here λ = m and μ = p.

Now let $P = x^{\lambda+1}(a + bx^n)^{\mu+1} = x^{m+1}(a + bx^n)^{p+1}$

$$\therefore \frac{dP}{dx} = x^{m+1}(p + 1)(a + bx^n)^p nb x^{n-1}$$

$$+ (m + 1)x^m (a + bx^n)^{p+1}$$

$$= bn(p + 1)x^{m+n}(a + bx^n)^p + (m + 1)x^m (a + bx^n)^p (a + bx^n)^p$$

(Note)

$$= bn(p + 1)x^{m+n}(a + bx^n)^p + (m + 1)ax^m (a + bx^n)^p$$

$$+ b(m + 1)x^{m+n}(a + bx^n)$$

$$= b(np + n + m + 1)x^{m+n}(a + bx^n)^p + (m + 1)ax^m(a + bx^n)^p,$$

which is linear combination of the two expressions whose integrals are to be connected.

Integrating both sides, we have

$$P = b(np + n + m + 1)\int x^{m+n}(a + bx^n)^p dx$$

$$+ (m + 1)a\int x^m (a + bx^n)^p dx$$

Now putting the value of P, dividing a (m + 1) and transposing suitably, we get $\int x^m (a + bx^n)^p dx$.

$$= \frac{x^{m+1}(a + bx^n)^{p+1}}{a(m + 1)} - \frac{b(np + n + m + 1)}{a(m + 1)}\int x^{m+1}(a + bx^n)^p dx$$

(iv) To connect $\int x^m (a + bx^n)^p dx$ with $\int x^m (a + bx^n)^{p-1} dx$

Here $\lambda = m$ and $\mu = p$, λ being the lesser index of x, and μ being the lesser indes of $(a + bx^n)$ in both the integrals.

Now let $P = x^{\lambda+1}(a + bx^n)^{m+1} = x^{m+1}(a + bx^n)^{p+1}$

$$\therefore \frac{dP}{dx} = x^{m+1}(p+1)(a+bx^n)^p \cdot nbx^{n-1}$$
$$+ (m + 1)x^m(a + bx^n)^{p+1}$$

$= n(p + 1)x^m(a + bx^n)^p\, bx^n + (m + 1)x^m(a + bx^n)^{p+1}$ (Note)

$= n(p + 1)x^m(a + bx^n)^p (a + bx^n - a) + (m + 1)x^m(a + bx^n)^{p+1}$

$+ (m + 1)x^m(a + bx^n)^{p+1}$

$= (np + n + m + 1)x^m(a + bx^n)^{p+1} - an(p + 1)x^m(a + bx^n)^p$

i.e. $\left(\frac{dP}{dx}\right)$ is a linear combination of the two expressions whose integrals are to be connected.

Integrating both the sides, we have

$$P = \{n(p+1) + m + 1\}\int x^m (a + bx^n)^{p+1} dx$$
$$- an(p+1)\int x^m (a + bx^n)^{p+1} dx$$

Now putting the value of P, dividing by $\{an(p + 1)\}$ and transposing suitably, we get $\int x^m (a + bx^n)^p dx$

$$= -\frac{x^{m+1}(a+bx^n)^{p+1}}{an(p+1)} + \frac{(np+n+m+1)}{an(n+1)}\int x^m (a+bx^n)^{p+1} dx$$

(v) To connect $\int x^m (a + bx^n)^p dx$ with $\int x^{m-n} (a + bx^n)^{p-1} dx$

Here $\lambda = m - n$ and $\mu = p$, [as also in case (i)]

$\therefore P = x^{\lambda+1}(a + bx^n)^{\mu+1} = x^{m-n+1}(a + bx^n)^{p+1}$

$$\therefore \frac{dP}{dx} = x^{m-n+1}(p+1)(a+bx^n)^p \cdot nbx^{n-1}$$
$$+ (m - n + 1)x^{m-n}(a + bx^n)^{p+1}$$

$= bn(p + 1)x^m(a + bx^n)^p + (m - n + 1)x^{m-n}(a + bx^n)^{p+1}$

= a linear combination of the two expresions whose integrals are to be connected.

$\therefore$ integrating both sides, we have

$$P = bn(p+1)\int x^m (a+bx^n)^p\,dx + (m-n+1)\int x^{m-n}(a+bx^n)^{p+1}dx$$

or $$bn(p+1)\int x^m(a+bx^n)^p\,dx = P - (m-n+1)\int x^{m-n}(a+bx^n)^{p+1}dx$$

Now putting the value of P and dividing by bn(p + 1), we get

$$\int x^m(a+bx^n)^p\,dx = \frac{x^{m-n+1}(a+bx^n)^{p+1}}{bn(p+1)} - \frac{(m-n+1)}{bn(p+1)}\int x^{m-n}(a+bx^n)^{p+1}dx$$

(vi) To connect $\int x^m(a+bx^n)^p\,dx$ with $\int x^{m+n}(a+bx^n)^{p-1}dx$

Here $\lambda = m$ and $\mu = p - 1$.

$\therefore\ P = x^{\lambda+1}(a+bx^n)^{\mu+1} = x^{m+1}(a+bx^n)^p$

And

$$\frac{dP}{dx} = x^{m+1}\cdot p(a+bx^n)^{p-1}bnx^{n-1} + (m+1)x^m(a+bx^n)^p$$

$$= bpnx^{m+n}(a+bx^n)^{p-1} + (m+1)x^m(a+bx^n)^p$$

Thus $\left(\frac{dP}{dx}\right)$ is expressed as a linear combination of the two expressions whose integrals are to be connected.

Integrating both sides, we have

$$P = bpn\int x^{m+n}(a+bx^n)^{p-1}dx + (m+1)\int x^m(a+bx^n)^p\,dx$$

or $$(m+1)\int x^m(a+bx^n)^p dx = P - bnp\int x^{m+n}(a+bx^n)^{p-1}dx$$

Now putting the value of P and dividing by (m + 1), we get

$$\int x^m(a+bx^n)^p\,dx = \frac{x^{m+1}(a+bx^n)^p}{(m+1)} - \frac{bnp}{(m+1)}\int x^{m+n}(a+bx^n)^{p-1}dx.$$

3.2. REDUCTION FORMULA FOR $\int \frac{dx}{(x^2+a^2)^n}$, WHERE n IS POSITIVE

Let $I_n = \int \frac{1}{(x^2+a^2)^n}dx$. To form a reduction formula for I_n, we shall integrate by parts $\int \frac{1}{(x^2+a^2)^{n-1}}\,dx$, taking unity as the second functon. Thus,

$$\int \frac{1}{(x^2+a^2)^{n-1}} \cdot 1\,dx = \frac{x}{(x^2+a^2)^{n-1}} - \int x \cdot \frac{-(n-1)}{(x^2+a^2)^n} \cdot 2x\,dx$$

or $$I_{n-1} = \frac{x}{(x^2+a^2)^{n-1}} + 2(n-1)\int \frac{x^2}{(x^2+a^2)^n}\,dx$$

$$= \frac{x}{(x^2+a^2)^{n-1}} + 2(n-1)\int \frac{(x^2+a^2)-a^2}{(x^2+a^2)^n}\,dx \qquad \text{(Note)}$$

$$= \frac{x}{(x^2+a^2)^{n-1}} + 2(n-1)\int \frac{1}{(x^2+a^2)^{n-1}}\,dx$$

$$-2(n-1)a^2 \int \frac{1}{(x^2+a^2)^n}\,dx$$

$$= \frac{x}{(x^2+a^2)^{n-1}} + 2(n-1)I_{n-1} - 2(n-1)a^2 I_n$$

$$\therefore\ 2(n-1)a^2 I_n = \frac{x}{(x^2+a^2)^{n-1}} + (2n-2-1)I_{n-1}$$

or $$I_n = \frac{x}{2a^2(n-1)(x^2+a^2)^{n-1}} + \frac{2n-3}{2a^2(n-1)} I_{n'},$$

which is the required reduction formula.

3.3. REDUCTION FORMULA FOR $\int x^m \sqrt{(2ax - x^2)}\,dx$; m BEING A POSITIVE INTEGER.

Let $I_m = \int x^m \sqrt{(2ax - x^2)}\,dx = \int x^{m+1/2} \sqrt{(2a - x)}\,dx$

Integratig by parts taking $\sqrt{(2a-x)}$ as the 2nd function, we have

$$I_m = x^{m+1/2} \frac{(2a-x)^{3/2}}{\left(\frac{3}{2}\right)\cdot(-1)} - \int \left(m + \frac{1}{2}\right) x^{m-1/2} \frac{(2a-x)^{3/2}}{\left(\frac{3}{2}\right)\cdot(-1)}\,dx$$

$$= -\frac{2}{3} x^{m-1} x^{3/2} (2a-x)^{3/2}$$

$$+ \frac{2m+1}{3} \int x^{m-1/2} (2a-x)\sqrt{(2a-x)}\,dx$$

$$= -\frac{2}{3}x^{m-1}(2ax - x^2)^{3/2} + \frac{2m+1}{3}\int 2ax^{m-1/2}\sqrt{(2a - x)}\,dx$$

$$- \frac{2m+1}{3}\int x^{m-1/2}\cdot x\sqrt{(2a - x)}\,dx$$

$$= -\frac{2}{3}x^{m-1}(2ax - x^2)^{3/2}$$

$$+ \frac{(2m+1)2a}{3}\int x^{m-1}x^{1/2}\sqrt{(2a - x)^{1/2}}\,dx$$

$$- \frac{2m+1}{3}\int x^{m-1/2}\cdot x^{1/2}x^{1/2}(2a - x)^{1/2}\,dx$$

$$= -\frac{2}{3}x^{m-1}(2ax - x^2)^{3/2} + \frac{2(2m+1)a}{3}\int x^{m-1}\sqrt{(2ax - x^2)}\,dx$$

$$- \frac{2m+1}{3}\int x^m\sqrt{(2ax - x^2)}\,dx$$

$$= -\frac{2}{3}x^{m-1}(2ax - x^2)^{3/2} + \frac{2(2m+1)}{3}2I_{m-1} - \frac{2m+1}{3}I_m.$$

Transposing the last term to the left, we have

$$\left(1 + \frac{2m+1}{3}\right)I_m = -\frac{2}{3}x^{m-1}(2ax - x^2)^{3/2} + \frac{2(2m+1)a}{3}I_{m-1}$$

or $$\frac{2(m+2)}{3}I_m = -\frac{2}{3}x^{m-1}(2ax - x^2)^{3/2} + \frac{2(2m+1)a}{3}I_{m-1}$$

or $$I_m = -\frac{x^{m-1}(2ax - x^2)^{3/2}}{m+2} + \frac{(2m+1)a}{n+2}I_{m-1},$$

which is the required reduction formula.

3.4. REDUCTION FORMULAE FOR

(a) $\int e^{mx}x^n dx$ and (b) $\int \frac{e^{mx}}{x^n}dx$ (n > 0).

(a) We have $\int e^{mx}x^n dx = x^n\frac{e^{mx}}{m} - \int nx^{n-1}\frac{e^{mx}}{m}dx$

integrating by parts taking e^{mx} as the 2nd function

$$= \frac{x^n e^{mx}}{m} - \frac{n}{m}\int x^{n-1}e^{mx}dx$$

which is the required reduction formula.

By repeated application of this formula the integral shall ultimately reduce to $\int x^0 e^{mx}dx$ and we have

$$\int x^0 e^{mx} dx = \int e^{mx} dx = \frac{e^{mx}}{m}$$

(b) We have $\int \frac{e^{mx}}{x^n} dx = e^{mx} \cdot x^{-n} dx$

$$= e^{mx} \cdot \frac{x^{-n+1}}{-n+1} - \int \frac{x^{-n+1}}{-n+1} \cdot me^{mx} dx$$

integrating by parts regarding x^{-n} as the second function

$$= \frac{-e^{mx}}{(n-1)x^{n-1}} + \frac{m}{n-1} \int \frac{e^{mx}}{x^{n-1}} dx,$$

which is the required reduction formula.

3.5. REDUCTION FORMULAE FOR

(a) $\int a^x x^n dx$ **and (b)** $\int \left(\frac{a^x}{x^n}\right) dx$

(a) Integrate by parts taking α^x as the second function. The required reduction formula is

$$\int a^x x^n dx = \frac{\alpha^x x^n}{\log a} - \frac{n}{\log a} \int x^{n-1} \alpha^x dx$$

(b) We have $\int (\alpha^x / x^n) dx = \int \alpha^x x^{-n} dx$.

Now integrate by parts taking x^{-n} as the 2nd function.

3.6. REDUCTION FORMULA FOR $\int x^m (\log x)^n \, dx$

Integrating by parts regarding x^m as the 2nd function, we get

$\int x^m (\log x)^n dx = (\log x)^n \cdot \frac{x^{m+1}}{m+1} - \frac{n}{m+1} \int x^m (\log x)^{n-1} dx$, which is the required reduction formula.

MISCELLANEOUS EXAMPLES

Example 1:

If $u_n = \int x^n (a^2 - x^2)^{1/2} dx$, *prove that*

$$u_n = -\frac{x^{n-1}(a^2 - x^2)^{3/2}}{n+2} + \frac{n-1}{n+2} a^2 u_{n-2}$$

Hence evaluate $\int_0^a x^4 \sqrt{(a^2 - x^2)}\, dx$.

Solution:

We have $u_n = \int x^n (a^2 - x^2)^{1/2} dx$

$$= -\frac{1}{2}\int x^{n-1} \cdot \{(a^2 - x^2)^{1/2} \cdot (-2x)\}\, dx$$

$$= -\frac{1}{2}x^{n-1}\left[\frac{2}{3}(a^2 - x^2)^{3/2}\right] + \frac{1}{3}(n-1)\int x^{n-2}(a^2 - x^2)^{3/2} dx,$$

integrating by parts taking $(a^2 - x^2)^{1/2}(-2x)$ as the second function

$$= -\frac{1}{3}x^{n-1}(a^2 - x^2)^{3/2}$$

$$+ \frac{1}{3}(n-1)\int x^{n-2}(a^2 - x^2)^{1/2}(a^2 - x^2)dx \quad \text{(Note)}$$

$$= -\frac{1}{3}x^{n-1}(a^2 - x^2)^{3/2} + \frac{1}{3}(n-1)u_{n-2} + \frac{1}{3}(n-1)u_n$$

Transposing the last term to the left, we have

$$\left\{1 + \frac{1}{3}(n-1)\right\}u_n = -\frac{1}{3}x^{n-1}(a^2 - x^2)^{3/2} + \frac{1}{3}(n-1)a^2 u_{n-2}$$

$$\text{or } u_n = -\frac{x^{n-1}(a^2 - x^2)^{3/2}}{(n+2)} + \frac{n-1}{n+2}a\,u_{n-2} \qquad \text{...(1)}$$

Taking limits x = 0 to x = a in (1), we have

$$\int_0^a x^n \sqrt{(a^2 - x^2)}dx$$

$$= -\left[\frac{x^{n-1}(a^2 - x^2)^{3/2}}{(n+2)}\right]_0^a + \frac{(n+1)a^2}{(n+2)}\int_0^a x^{n-2}\sqrt{(a^2 - x^2)}\,dx$$

$$= 0 + \frac{n-1}{n+2}a^2\int_0^a x^{n-2}(a^2 - x^2)^{1/2}dx$$

Putting n = 4 in (2), we have

$$\int_0^a x^4 (a^2 - x^2)^{1/2}\, dx = \frac{4-1}{4+2}a^2 \int_0^a x^2\, (a^2 - x^2)^{1/2} dx$$

$$= \frac{3a^2}{6} \cdot \frac{2-1}{2+1}a^2 \int_0^a x^0 (a^2 - x^2)^{1/2} dx$$

again applying (2) by taking n =

$$= \frac{1}{8}a^4\left[\frac{1}{2}x\sqrt{(a^2 - x^2)} + \frac{1}{2}a^2 \sin^{-1}\frac{x}{a}\right]_0^a$$

$$= \frac{1}{8}a^4 \cdot \frac{1}{2}a^2 \cdot \frac{1}{2}\pi = \frac{\pi a^6}{32}.$$

Example 2:

Evaluate $\int_0^\infty e^{-x}x^n dx$, *n being a positive integer.*

Solution:

Integrating by parts regarding e^{-x} as the 2nd function, we get get

$$\int_0^\infty e^{-x}x^n dx = \left[-x^n e^{-x}\right]_0^\infty + \int_0^\infty ne^{-x}x^{n-1}dx$$

Now $\lim_{x\to\infty} x^n e^{-x} = \lim_{x\to\infty}\frac{x^n}{e^x}$, which is of the form $\frac{\infty}{\infty}$.

$\therefore$ differentiating the numerator and the denominator separately, we get

$$\lim_{x\to\infty}\frac{x^n}{e^x} = \lim_{x\to\infty}\frac{nx^{n-1}}{e^x} = \ldots = \lim_{x\to\infty}\frac{n(n-1)\ldots 1}{e^x} = 0$$

Hence $\left[-x^n e^{-x}\right]_0^\infty = -\lim_{x\to\infty} x^n e^{-x} - 0 = 0 - 0 = 0$

Therefore $\int_0^\infty e^{-x}x^n dx = n\int_0^\infty e^{-x}x^{n-1}dx$ ...(1)

Now applying the reduction formula (1) repeatedly, we get

$$\int_0^\infty e^{-x}x^n dx = n(n-1)(n-2)\ldots 2.1\int_0^\infty e^{-x}x^0 dx$$

$$= n!\int_0^\infty e^{-x}dx = n!\left[-e^{-x}\right]_0^\infty = n!\left[-\frac{1}{e^x}\right]_0^\infty = (n!).1 = n!$$

Example 3:

Show that $\int_0^\infty e^{-ax}x^x dx = \frac{n!}{a^{n+1}}$, *where a is a positive quantity and n is a positive integer.*

Solution:

Interating by parts regarding e^{-ax} as the 2nd function, we get

$$\int_0^\infty e^{-ax}x^n dx = \left[x^n\left(\frac{e^{-ax}}{-a}\right)\right]_0^\infty - \int_0^\infty nx^{n-1}\left(\frac{e^{-ax}}{-a}\right)dx,$$

$$= 0 + \frac{n}{a}\int_0^\infty e^{-ax}x^{n-1}dx,$$

Thus $\displaystyle\int_0^\infty e^{-ax}x^n dx = \frac{n}{a}\int_0^\infty e^{-ax}x^{n-1}dx$ ...(1)

Now by the repeated application of the reduction formula (1), we have ultimately

$$\int_0^\infty e^{-ax}x^n dx = \frac{n!}{a^n}\int_0^\infty e^{-ax}x^0 dx = \frac{n!}{a^n}\left[-\frac{1}{ae^{ax}}\right]_0^\infty = \frac{n!}{a^n}\cdot\frac{1}{a} = \frac{n!}{a^{n+1}}$$

Example 4:

Evaluate $\int_0^1 x^m(\log x)^n\,dx$ *when* $m \ge 0$ *and n is an integer* ≥ 0

Solution:

Let $I_{m,n} = \int_0^1 x^m(\log x)^n\,dx$

Integrating by parts taking x^m as the second function, we have

$$I_{m,n} = \left[\frac{x^{m+1}(\log x)^n}{m+1}\right]_0^1 - \frac{n}{m+1}\int_0^1 x^m(\log x)^{n-1}dx$$

$$= \frac{1}{m+1}\left[1^{m+1}(\log 1)^n - \lim_{x\to 0} x^{m+1}(\log x)^n\right]$$

$$- \frac{n}{m+1}\int_0^1 x^m(\log x)^{n-1}dx$$

But $\log 1 = 0$. Also $\displaystyle\lim_{x\to 0} x^{m+1}(\log x)^n = \lim_{x\to 0}\frac{(\log x)^n}{x^{-(m+1)}}$

$$= \lim_{x\to 0}\frac{n(\log x)^{n-1}\cdot\left(\frac{1}{x}\right)}{-(m+1)x^{-(m+2)}} = \lim_{x\to 0}\left(-\frac{n}{m+1}\right)\frac{(\log x)^{n-1}}{x^{-(m+1)}}$$

Proceeding in this way, we ultimately have

$$\lim_{x\to 0} x^{m+1}(\log x)^n = (\text{some number}) \times \lim_{x\to 0}\frac{1}{x^{-(m+1)}}$$

$$= (\text{some number}) \times \lim_{x\to 0} x^{m+1} = 0.$$

$\therefore\ \displaystyle I_{m,n} = \frac{n}{(m+1)}I_{m,n-1}$ ...(1)

$$= -\frac{n}{(m+1)}\cdot\left(-\frac{n-1}{m+1}\right)I_{m,n-2}, \text{ applying (1)}$$

$$= (-1)^2 \frac{n(n-1)}{(m+1)^2} I_{m, n-1} = \ldots = \ldots$$

Similarly by successive application of (1), we have ultimately

$$I_{m, n} = (-1)^n \frac{n(n-1)(n-2)\ldots 2.1}{(m+1)^n} I_{m, 0}$$

But $I_{m, 0} = \int_0^1 x^m (\log x)^0 \, dx = \int_0^1 x^m dx$

$$= \left[\frac{x^{m+1}}{m+1}\right]_0^1 = \frac{1}{m+1}$$

$$\therefore\ I_{m, n} = (-1)^n \cdot \frac{n!}{(m+1)^n} \cdot \frac{1}{(m+1)} = (-1)^n \frac{n!}{(m+1)^{n+1}}$$

Example 5:

If I_n denotes $\int_0^1 x^p (1 - x^q)^n \, dx$, where p, q and n are positive, prove that $(nq + p + 1)\, I_n = nq\, I_{n-1}$.

Hence evaluate I_n when n is a positive integer.

Solution:

Here we have to connect

$$\int_0^1 x^p (1 - x^q)^n \, dx \text{ with } \int_0^1 x^p (1 - x^q)^{n-1} \, dx$$

$\therefore$ Here λ = lesser index of x = p;

μ = lesser index of $(1 - x^q) = n - 1$

$\therefore\ P = x^{\lambda+1}(1 + x^q)^{\mu+1} = x^{p+1}(1 - x^q)^n$

Hence

$$\frac{dP}{dx} = (p+1)x^p (1 - x^q)^n + x^{p+1} \cdot n(1 - x^q)^{n-1} \cdot (-qx^{q-1})$$

$= (p + 1)x^p (1 - x^q)^n + nqx^p (1 - x^q)^{n-1} \cdot (-x^q)$

$= (p + 1)x^p (1 - x^q)^n + nqx^p (1 - x^q)^{n-1} \cdot \{(1 - x^q) - 1\}$ (Note)

$= (p + 1)x^p (1 - x^q)^n + nqx^p (1 - x^q)^n - nqx^p (1 - x^q)^{n-1}$

$= (p + 1 + nq)x^p (1 - x^q)^n - npx^p (1 - x^q)^{n-1}$.

Thus $\left(\frac{dP}{dx}\right)$ is expressed as a linear combination of the two expressions whose integrals are to be connected. Therefore, integrating both sides, we have

$$P=(p+1+nq)\int x^p(1-x^q)^n\,dx-nq\int x^p(1-x^q)^{n-1}\,dx$$

$$\therefore\ (p+1+nq)\int_0^1 x^p(1-x^q)^n\,dx$$

$$=\left[x^{p+1}(1-x^q)^n\right]_0^1+nq\int_0^1 x^p(1-x^q)^{n-1}\,dx,$$

putting the value of P, transposing and also putting the limits of integration

$$=0+nq\int_0^1 x^p(1-x^q)^{n-1}\,dx.$$

Thus $(qn+p+1)I_n=nq\,I_{n-1}$ ...(1)

or $I_n=\dfrac{nq}{qn+p+1}\cdot I_{n-1}=\dfrac{nq}{qn+p+1}\cdot\left[\dfrac{(n-1)q}{\{(n-1)q+p+1\}}I_{n-2}\right].$

putting $(n-1)$ for n in (1) to get I_{n-1} in terms of I_{n-2}. Proceeding similarly by successive reduction, we have finally

$$I_n=\frac{nq}{nq+p+1}\cdot\frac{(n-1)q}{(n-1)q+p+1}\cdots\frac{q}{q+p+1}I_0$$

$$\text{But } I_0=\int_0^1 x^p(1-x^q)^0\,dx=\int_0^1 x^p dx=\left[\frac{x^{p+1}}{p+1}\right]_0^1=\frac{1}{p+1}$$

$$\therefore\ I_n=\frac{nq}{nq+p+1}\cdot\frac{(n-1)q}{(n-1)q+p+1}\cdots\frac{q}{q+p+1}\cdot\frac{1}{p+1}.$$

Example 6:

If I_n denotes $\int_0^a (a^2-x^2)^n\,dx$, and $n>0$, prove that

$$I_n=\frac{2na^2}{2n+1}I_{n-1}.$$

Hence evaluate $\int_0^a (a^2-x^2)^3\,dx$.

Solution:

We have $I_n=\int_0^a (a^2-x^2)^n\cdot 1\,dx$ (Note)

$$=\left[(a^2-x^2)^n\cdot x\right]_0^a-\int_0^a n(a^2-x^2)^{n-1}(-2x)\cdot x\,dx,$$

integrating by parts taking unity as the second function

$$= 0 + 2n\int_0^a (a^2 - x^2)^{n-1} x^2 dx, \qquad [\because n > 1]$$

$$= -2n\int_0^a (a^2 - x^2)^{n-1} \cdot \{(a^2 - x^2) - a^2\}\, dx,$$

$$[\because x^2 = -\{(a^2 - x^2) - a^2\}]$$

$$= -2n\int_0^a (a^2 - x^2)^n dx + 2na^2 \int_0^a (a^2 - x^2)^{n-1} dx$$

$$= -2nI_n + 2na^2 I_{n-1}.$$

$$\therefore (1 + 2n) I_n = 2na^2 I_{n-1}$$

or $$I_n = \frac{2na^2}{2n+1} I_{n-1} \qquad ...(1)$$

$\therefore I_3 = \frac{6}{7} a^2 I_2$, putting n = 3 in (1)

$= \frac{6}{7} a^2 \cdot \left[\frac{4}{5} a^2 I_1\right]$, putting n = 2 in (1) to get I_2 in terms of

$I_1 = \frac{24}{35} a^4 I_1$.

Thus $$I_3 = \int_0^a (a^2 - x^2)^3 dx = \frac{24a^4}{35}\int_0^a (a^2 - x^2) dx$$

$$= \frac{24a^4}{35}\left[a^2 x - \frac{x^3}{3}\right]_0^a = \frac{24a^4}{35}\left[a^3 - \frac{a^3}{3}\right] = \frac{16a^7}{35}.$$

Example 7:

Prove the reduction formula

$$\int (a^2 + x^2)^{n/2} dx = \frac{x(a^2 + x^2)^{n/2}}{(n+1)} + \frac{na^2}{(n+1)} \int (a^2 + x^2)^{(n/2)-1} dx$$

Hence evaluate $\int (x^2 + a^2)^{5/2} dx$.

Solution:

We have $\int (a^2 + x^2)^{n/2} dx = \int (a^2 + x^2)^{n/2} \cdot 1\, dx$ (Note)

$$= (a^2 + x^2)^{n/2} x - \int \frac{1}{2} n(a^2 + x^2)^{(n/2)-1} 2x \cdot x\, dx$$

$$= x(a^2 + x^2)^{n/2} - n\int (a^2 + x^2)^{(n/2)-1} \{(a^2 + x^2) - a^2\}\, dx$$

(Note)

$$= x(a^2+x^2)^{n/2} - n\int (a^2+x^2)^{n/2}\,dx + na^2\int (a^2+x^2)^{(n/2)-1}\,dx$$

$$\therefore\ (1+n)\int (a^2+x^2)^{n/2}\,dx = x(a^2+x^2)^{n/2} + na^2\int (a^2+x^2)^{(n/2)-1}\,dx$$

or $$\int (a^2+x^2)^{n/2}\,dx = \frac{x(a^2+x^2)^{n/2}}{(n+1)} + \frac{na^2}{(n+1)}\int (a^2+x^2)^{(n/2)-1}\,dx \quad ...(1)$$

is the required reduction formula

Now to evaluate $\int (a^2+x^2)^{5/2}\,dx$, putting n = 5 in (1), we get

$$\int (a^2+x^2)^{5/2}\,dx = \frac{x(a^2+x^2)^{5/2}}{6} + \frac{5a^2}{6}\int (a^2+x^2)^{3/2}\,dx$$

$$= \frac{x(a^2+x^2)^{5/2}}{6} + \frac{5a^2}{6}\left[\frac{x(a^2+x^2)^{3/2}}{4} + \frac{3a^2}{4}\int (a^2+x^2)^{1/2}\,dx\right]$$

$$= \frac{x(a^2+x^2)^{5/2}}{6} + \frac{5a^2}{24}x(a^2+x^2)^{3/2} + \frac{5a^2}{16}\left[x\sqrt{(a^2+x^2)} + a^2\sin^{-1}\frac{x}{a}\right]$$

Example 8:

Evaluate $\int_0^\infty \frac{1}{(a^2+x^2)^4}\,dx$

Solution:

Let $I_n = \int_0^\infty \frac{1}{(a^2+x^2)^n}\,dx,$

where n is a + ve integer ≥ 2.

To form a reduction formula for I_n, we shall integrate by parts $\int_0^\infty \frac{1}{(a^2+x^2)^{n-1}}\,dx$, taking 1 as the 2nd function. Thus

$$\int_0^\infty \frac{1}{(a^2+x^2)^{n-1}}\cdot 1\,dx = \left[\frac{x}{(a^2+x^2)^{n-1}}\right]_0^\infty - \int_0^\infty x\,\frac{-(n-1)}{(a^2+x^2)^n}\cdot 2\,dx$$

or $I_{n-1} = \left[\lim_{x\to\infty} \frac{x}{(a^2+x^2)^{n-1}} - 0\right] + 2(n-1)\int_0^\infty \frac{x^2}{(a^2+x^2)^n}\,dx$

$$= 0 + 2(n-1)\int_0^\infty \frac{(a^2+x^2)-a^2}{(a^2+x^2)^n}\,dx$$

$$\left[\because \lim_{x\to\infty} \frac{x}{(a^2+x^2)^{n-1}} = 0 \text{ if } n \ge 2\right]$$

$$= 2(n-1)\int_0^\infty \frac{dx}{(a^2+x^2)^{n-1}} - 2(n-1)a^2\int_0^\infty \frac{1}{(a^2+x^2)^n}\,dx$$

$= 2(n-1)\, I_{n-1} - 2(n-1)a^2 I_n.$

$\therefore\ 2(n-1)a^2 I_n = \{2(n-1)-1\}I_{n-1}$

or $$I_n = \frac{2n-3}{2a^2(n-1)} I_{n-1}, \quad ...(1)$$

is the reduction formula for I_n.

Now putting $n = 4$ in (1), we get

$$I_4 = \frac{5}{2a^2\cdot 3} I_3 = \frac{5}{6a^2}\cdot\frac{3}{2a^2\cdot 2} I_2,$$

putting $n = 3$ in (1) to get I_3 in terms of I_2

$$= \frac{5}{8a^4}\cdot\frac{1}{2a^2\cdot 1} I_1 = \frac{5}{16a^6}\int_0^\infty \frac{1}{(a^2+x^2)}\,dx$$

$$= \frac{5}{16a^6}\frac{1}{a}\left[\tan^{-1}\frac{x}{a}\right]_0^\infty = \frac{5}{16a^7}[\tan^{-1}\infty - \tan^{-1}0]$$

$$= \frac{5}{16a^7}\cdot\frac{\pi}{2} = \frac{5\pi}{32a^7}.$$

Aliter. Let $I = \int_0^\infty \frac{1}{(a^2+x^2)^4}\,dx$

Put $x = a\tan\theta$, so that $dx = a\sec^2\theta\, d\theta$.

The limits for θ are from 0 to $\pi/2$.

$$\therefore\ I = \int_0^{\pi/2} \frac{1}{(a^2\sec^2\theta)^4}\cdot a\sec^2\theta\, d\theta = \frac{1}{a^7}\int_0^{\pi/2}\cos^6\theta\, d\theta$$

$$= \frac{1}{a^7}\cdot\frac{5\cdot 3\cdot 1}{6\cdot 4\cdot 2}\cdot\frac{\pi}{2}, \text{ by Walli's formula } = \frac{5\pi}{32a^7}.$$

Examle 9:

Prove that $\int_0^\infty \frac{dx}{(1+x^2)^4} = \frac{5\pi}{32}$

Solution:

Do your self. Put x = tan θ.

Example 10:

If $I_m = \int_0^{2a} x^m \sqrt{(2ax - x^2)}\, dx$, *prove that*

$2^m m!.\ (m + 2)\ !\ I_m = a^{m+2}\ (2m + 1)\ !\ \pi$

Hence or otherwise evaluate $\int_0^{2a} x^3 \sqrt{(2ax - x^2)}\, dx$.

Solution:

Proceeding as in taking limts from 0 to 2a, we get

$$I_m = -\left[\frac{x^{m-1}\ (2ax - x^2)^{3/2}}{m+2}\right]_0^{2a} + \frac{(2m+1)a}{m+2} I_{m-1}$$

$$\text{or } I_m = 0 + \frac{(2m+1)a}{m+2} I_{m-1} = \frac{(2m+1)a}{m+2} I_{m-1}$$

$$\therefore I_m = \frac{(2m+1)a}{m+2} . \frac{(2m-1)a}{m+1} I_{m-2}$$

replacing m by m – 1 in (1) to get I_{m-1} in terms of I_{m-2}

$$= \frac{(2m+1)a}{m+2} . \frac{(2m-1)a}{m+1} . \frac{(2m-3)a}{m} I_{m-3}, \text{ and so on}$$

Thus aplying the reducting formula (1) successively, we have fina

$$I_m = \frac{(2m+1)a}{m+2} . \frac{(2m-1)a}{m+1} . \frac{(2m-3)a}{m} \ldots \frac{5a}{4} . \frac{3a}{3} I_0,$$

where $I_0 = \int_0^{2a} x^0 \sqrt{(2ax - x^2)}\, dx = \int_0^{2a} \sqrt{x} . \sqrt{(2a - x)}\, dx$

Put x = 2a sin²θ,

so that dx = 4a sinθ cosθ dθ,

and the new limits are from θ = 0 to θ = π/2.

$$\therefore I_0 = \int_0^{\pi/2} \sqrt{(2a)}\ \sin\theta . \sqrt{(2a)}\ \cos\theta . 4a \sin\theta \cos\theta\ d\theta$$

$$= 8a^2 \int_0^{\pi/2} \sin^2\theta \cos^2\theta \, d\theta = 8a^2 \cdot \frac{\Gamma\frac{3}{2} \cdot \Gamma\frac{3}{2}}{2\Gamma 3}$$

$$= 8a^2 \frac{\frac{1}{2}\sqrt{\pi} \cdot \frac{1}{2}\sqrt{\pi}}{2 \cdot 2 \cdot 1} = \frac{\pi a^2}{2}.$$

$$\therefore \; I_m = \frac{a^m (2m+1)(2m-1)(2m-3) \ldots 5 \cdot 3}{(m+2)(m+1)\, m \ldots 4 \cdot 3} \cdot \frac{\pi a^2}{2}$$

$$= \frac{\pi a^{m+2}}{(m+2)!} \cdot (2m+1)(2m-1)(2m-3) \ldots 5 \cdot 3.$$

Multiplying the numerator and the denominator by $2m \cdot (2m - 2) \ldots 4.2$, we get

$$I_m = \frac{\pi a^{m+2}}{(m+2)!} \cdot \frac{(2m+1)2m\,(2m-1)(2m-2) \ldots 5 \cdot 4 \cdot 3 \cdot 2 \cdot 1}{2m \cdot (2m-2) \ldots 4 \cdot 2}$$

$$= \frac{\pi a^{m+2}}{(m+2)!} \cdot \frac{(2m+1)!}{2^m [m \cdot (m-1) \ldots 2 + 1]} = \frac{\pi a^{m+2} \cdot (2m+1)!}{2^m \cdot (m)! \cdot (m+2)!}$$

or $2^m(m)!\ (m + 2)!\ I_m = a^{m+2} \cdot (2m + 1)!\pi.$ Proved.

Now let $I = \int_0^{2a} x^3 \sqrt{(2ax - x^2)}\, dx = \int_0^{2a} x^{3+1/2} \sqrt{(2a - x)}\, dx$

$$= \int_0^{2a} x^{7/2} \sqrt{(2a - x)}\, dx.$$

Put $x = 2a \sin^2\theta$,

so that $dx = 4a \sin\theta \cos\theta \, d\theta$,

and the new limits are $\theta = 0$ to $\theta = \pi/2$.

$$\therefore \; I = \int_0^{\pi/2} (2a)^4 \sin^7\theta \cdot \cos\theta \cdot 4a \sin\theta \cos\theta \, d\theta$$

$$= 64a^5 \int_0^{\pi/2} \sin^8\theta \cos^2\theta \, d\theta = 64a^5 \frac{\Gamma\frac{9}{2}\Gamma\frac{3}{2}}{2\Gamma 6} = \frac{7\pi a^5}{8}$$

Example 11:

If $I_n = \int x^n (a - x)^{1/2}\, dx$

$(2n + 3)\, I_n = 2an\, I_{n-1} - 2x^n(a - x)^{3/2}.$

Hence evaluate $\int_0^a x^2 \sqrt{(ax - x^2)}\, dx$

Solution:

We have $I_n = \int x^n (a-x)^{1/2}\,dx$

$$= x^n \cdot \left\{-\frac{2}{3}(a-x)^{3/2}\right\} - \int nx^{n-1} \cdot \left\{-\frac{2}{3}(a-x)^{3/2}\right\} dx$$

integrating by parts taking $(a-x)^{1/2}$ as the 2nd function

$$= -\frac{2}{3}x^n (a-x)^{3/2} + \frac{2}{3}n \int x^{n-1}(a-x)(a-x)^{1/2}\,dx \qquad \text{(Note)}$$

$$= -\frac{2}{3}x^n (a-x)^{3/2} + \frac{2}{3}na \int x^{n-1}(a-x)^{1/2}\,dx$$

$$-\frac{2}{3}n \int x^n (a-x)^{1/2}\,dx$$

$$= -\frac{2}{3}x^n (a-x)^{3/2} + \frac{2}{3}na\, I_{n-1} - \frac{2}{3}nI$$

Transposing the last term to the left, we get

$$\left(1+\frac{2}{3}\right) I_n = \frac{2}{3}na\, I_{n-1} - \frac{2}{3}x^n (a-x)^{3/2}$$

or $(2n+3)\, I_n = 2na\, I_{n-1} - 2x^n (a-x)^{3/2}$ **Proved.**

$$\text{or } I_n = \frac{2na}{(2n+3)} I_{n-1} - \frac{2x^n (a-x)^{3/2}}{(2n+3)} \qquad ...(1)$$

Taking limits $x = 0$ to $x = a$ in (1), we have

$$\int_0^a x^n (a-x)^{1/2}\,dx$$

$$= \frac{2na}{(2n+3)} \int_0^a x^{n-1}(a-x)^{1/2}\,dx - \left[\frac{2x^n (a-x)^{3/2}}{(2n+3)}\right]_0^a$$

$$= \frac{2na}{(2n+3)} \int_0^a x^{n-1}(a-x)^{1/2}\,dx - 0 \qquad ...(2)$$

To evaluate $\int_0^a x^2 \sqrt{(ax-x^2)}\,dx$,

we have $\int_0^a x^2 \sqrt{(ax-x^2)}\,dx$

$$= \int_0^a x^2 \cdot x^{1/2} \cdot \sqrt{(a-x)}\,dx = \int_0^a x^{5/2} (a-x)^{1/2}\,dx$$

$$= \frac{2a \cdot \frac{5}{2}}{2 \cdot \frac{5}{2} + 3} \int_0^a x^{3/2} (a-x)^{1/2}\,dx,$$

applying the reduction formula (2) by taking n = 5/2

$$= \frac{5a}{8} \cdot \frac{2a \cdot \frac{3}{2}}{2 \cdot \frac{3}{2} + 3} \int_0^a x^{1/2} (a - x)^{1/2} \, dx,$$

again aplying (2) by taking n = 3/2

$$= \frac{5a}{8} \cdot \frac{a}{2} \int_0^a x^{1/2} (a - x)^{1/2} \, dx$$

Now put $x = a \sin^2 \theta$,

so that $dx = 2a \sin \theta \cos \theta \, d\theta$,

and the new limits are $\theta = 0$ to $\theta = \pi/2$.

$$\therefore \int_0^a x^2 \sqrt{(ax - x^2)} \, dx$$

$$= \frac{5a^2}{16} \int_0^{\pi/2} a^{1/2} \sin\theta \cdot a^{1/2} \cos\theta \cdot 2a \sin\theta \cos\theta \, d\theta$$

$$= \frac{5a^2}{16} \cdot 2a^2 \int_0^{\pi/2} \sin^2\theta \cos^2\theta \, d\theta = \frac{5a^4}{8} \cdot \frac{\Gamma\frac{3}{2} \cdot \Gamma\frac{3}{2}}{2\Gamma 3}$$

by Gamma function

$$= \frac{5a^4}{8} \frac{\frac{1}{2}\sqrt{\pi} \cdot \frac{1}{2}\sqrt{\pi}}{2 \cdot 2 \cdot 1} = \frac{5\pi a^4}{128}.$$

Example 12:

Evaluate $\int_0^1 (\log x))0^4 \; x^m \, dx$

Solution:

Let $I_{m,n} = \int_0^1 x^m (\log x)^n \, dx$, where n is an integer ≥ 0. Then proceeding as in Ex. 13, we have

$$I_{m.n} = -\frac{n}{(m+1)} I_{m.n-1}$$

Putting m = 4 in (1), we get

$$I_{m,4} = \int_0^1 (\log x)^4 x^m dx = \frac{-4}{m+1} \int_0^1 x^m (\log x)^{4-1} dx$$

$$= -\frac{4}{m+1} I_{m,3}$$

Now by repeated application of (1), we have

$$I_{m,4} = \left(\frac{-4}{m+1}\right)\cdot\left(\frac{-3}{m+1}\right)\left(\frac{-2}{m+1}\right)\left(\frac{-1}{m+1}\right)\int_0^1 x^m (\log x)^0\, dx$$

$$= \left(\frac{-4}{m+1}\right)\cdot\left(\frac{-3}{m+1}\right)\cdot\left(\frac{-2}{m+1}\right)\cdot\left(\frac{-1}{m+1}\right)\int_0^1 x^m\, dx$$

$$= \frac{24}{(m+1)^4}\int_0^1 x^m dx = \frac{24}{(m+1)^4}\cdot\left(\frac{x^{m+1}}{m+1}\right)_0^1 = \frac{24}{(m+1)^5}$$

Example 13:

If m and n are positie integers, and $f(m, n) = \int_0^1 x^{n-1} (\log x)^m\, dx$, *proe that*

$$f(m, n) = -\left(\frac{m}{n}\right) f(m-1, n)$$

Deduce that $f(m, n) = (-1)^m \cdot \frac{m!}{n^{m+1}}$.

Solution:

Interating by parts regarding x^{n-1} as the 2nd function, we have

$$f(m, n) = \left[(\log x)^m \cdot \frac{x^n}{n}\right]_0^1 - \frac{m}{n}\int_0^1 (\log x)^{m-1} \cdot \frac{1}{x} \cdot x^n\, dx$$

$$= 0 - \frac{m}{n}\int_0^1 x^{n-1} (\log x)^{m-1}\, dx$$

$$\left[\because \lim_{x\to 0} x^n (\log x)^m = 0, \text{as shown in Ex. 13}\right]$$

or $f(m, n) = -\left(\frac{m}{n}\right) f(m-1, n)$ ...(1)

$$= (-1)^2 \left(\frac{m}{n}\right)\cdot\left(\frac{m-1}{n}\right) f(m-2, n), \text{ applying (1)}$$

$$= (-1)^3 \left(\frac{m}{n}\right)\left(\frac{m-1}{n}\right)\left(\frac{m-2}{n}\right) f(m-3, n), \text{ again applying (1)}$$

$$= (-1)^3 \frac{m(m-1)(m-2)}{n^3} f(m-3, n)$$

Similarly by successive application of (1), ultimately we have

$$f(m, n) = (-1)^m \frac{m(m-1)(m-2)\ldots 2.1}{n^m} f(0, n)$$

But $f(0, n) = \int_0^1 x^{n-1} (\log x)^0 \, dx$

$$= \int_0^1 x^{n-1} dx = \left[\frac{x^n}{n}\right]_0^1 = \frac{1}{n}$$

$$\therefore \; f(m, n) = (-1)^m \frac{m!}{n^m} \cdot \frac{1}{n} = (-1)^m \frac{m!}{n^{m+1}}$$

Example 14:

Find the reduction formula for $\int \left\{\frac{x^m}{(\log x)^n}\right\} dx$

Solution:

We have $\int \frac{x^m}{(\log x)^n} dx$

$$= \int x^{m+1} \left[\frac{1}{(\log x)^n} \cdot \frac{1}{x}\right] dx = \int x^{m+1} \cdot \left[(\log x)^{-n} \frac{1}{x}\right] dx \qquad \text{(Note)}$$

Now integrating by parts regarding x^{m+1} as the first function, we have

$$\int \frac{x^m dx}{(\log x)^n} = x^{m+1} \frac{(\log x)^{-n+1}}{-n+1} - \int (m+1) x^m \frac{(\log x)^{-n+1}}{-n+1} dx$$

$$= -\frac{x^{m+1}}{(n-1)(\log x)^{n-1}} + \frac{m+1}{n-1} \int \frac{x^m}{(\log x)^{n-1}} dx$$

which is the required reduction formula.

Example 15:

Evaluate $\int_0^\infty \frac{x}{(1+e^x)} dx.$

Solution:

We have $I = \int_0^\infty \frac{x \, dx}{(1+e^x)} = \int_0^\infty \frac{x \, dx}{e^x (1+e^{-x})}$

$$= \int_0^\infty x e^{-x} (1+e^{-x})^{-1} dx \qquad \text{(Note)}$$

On expanding $(1 + e^{-x})^{-1}$ by binomial theorem, we have

$(1 + e^{-x})^{-1} = 1 - e^{-x} + e^{-2x} - e^{-3x} + e^{-4x} - \ldots$

$$\therefore \; I = \int_0^\infty x e^{-x} [1 - e^{-x} + e^{-2x} - e^{-3x} + \ldots] \, dx$$

$$= \int_0^{\infty} x[e^{-x} - e^{-2x} + e^{-3x} + e^{-4x} + \ldots]\,dx \qquad \ldots(1)$$

Also $\int_0^{\infty} xe^{-nx}dx$

$$= \left[-\frac{xe^{-nx}}{n}\right]_0^{\infty} + \int_0^{\infty} \frac{e^{-nx}}{n}dx = 0 + \left[\frac{e^{-nx}}{-n^2}\right]_0^{\infty} = \frac{1}{n^2} \qquad \ldots(2)$$

Now applying (2) to each term of the R.H.S. of (1), we get

$$I = 1 - \frac{1}{2^2} + \frac{1}{3^2} - \frac{1}{4^2} + \frac{1}{5^2} - \ldots \infty$$

$= \frac{\pi^2}{12}$, from trigonometry.

Example 16:

Find a reduction formula for $\int x^m (1+x^2)^{n/2}\,dx$, where m and n are psotive integers.

Hence evaluate $\int x^5 (1+x^2)^{7/2}\,dx$

Solution:

We have $\int x^m (1+x^2)^{n/2}\,dx = \frac{1}{2}\int x^{m-1}(1+x^2)^{n/2} \cdot 2x\,dx$

$$= \frac{1}{2}\left[x^{m-1}\frac{(1+x^2)^{(n/2)+1}}{\frac{1}{2}n+1} - \frac{m-1}{\left(\frac{1}{2}n+1\right)}\int x^{m-2}(1+x^2)^{(n/2)+1}dx\right]$$

integrating by parts taking x^{m-1} as first function

$$= x^{m-1}\frac{(1+x^2)^{(n+2)/2}}{(n+2)} - \frac{m-1}{(n+2)}\int x^{m-2}(1+x^2)^{(n+2)/2}dx \qquad \ldots(1)$$

which is the required reduction formula.

Now to evaluate $\int x^5 (1+x^2)^{7/2}\,dx$, put m = 5, n = 7 in (1).

Then $\int x^5(1+x^2)^{7/2}dx = \frac{x^4(1+x^2)^{9/2}}{9} - \frac{4}{9}\int x^2(1+x^2)^{9/2}dx$

$$= \frac{x^4(1+x^2)^{9/2}}{9} - \frac{4}{9}\left[\frac{x^2(1+x^2)^{11/2}}{11} - \frac{2}{11}\int x(1+x^2)^{11/2}dx\right]$$

putting m = 3, n = 9 in (1)

$$= \frac{x^4(1+x^2)^{9/2}}{9} - \frac{4x^2(1+x^2)^{11/2}}{99} + \frac{4}{99}\int (1+x^2)^{11/2} \cdot 2x\, dx$$

$$= \frac{x^4(1+x^2)^{9/2}}{9} - \frac{4x^2(1+x^2)^{11/2}}{99} + \frac{4}{99}\frac{(1+x^2)^{13/2}}{(13/2)}$$

$$= \frac{1}{9}(1+x^2)^{9/2}\left[x^4 - \frac{4}{11}x^2(1+x^2) + \frac{8}{143}(1+x^2)^2\right].$$

Example 17:

If $I_{m,n} = \int \frac{x^m dx}{(1+x^2)^n}$, *prove that*

$$2(n-1)I_{m,n} = -x^{m-1}(x^2+1)^{1-n} + (m-1)I_{m-2,n-1}$$

Solution:

We have $I_{m,n} = \int \frac{x^m dx}{(1+x^2)^n}$

$$= \frac{1}{2}\int x^{m-1}\{(1+x^2)^{-n} \cdot 2x\}\, dx$$

$$\frac{1}{2}x^{m-1}\frac{(1+x^2)^{-n+1}}{(-n+)} - \frac{1}{2}\int (m-1)x^{m-2}\frac{(1+x^2)^{-n+1}}{(-n+1)}dx$$

integrating by parts taking $\{(1+x^2)^{-n}, 2x\}$ as the 2nd function

$$= \frac{-1}{2(n-1)}x^{m-1}(x^2+1)^{1/n} + \frac{(m-1)}{2(n-1)}\int \frac{x^{m-2}}{(1+x^2)^{n-1}}dx$$

Multiplying both sides by 2(n −1), we get

$2(n-1)I_{m,n} = -x^{m-1}(1+x^2)^{-n+1} + (m-1)\, I_{m-2,n-1}$,

which is the required reduction formula.

Example 18:

If $\phi(n) = \int_0^x \frac{x^n dx}{\sqrt{(x-1)}}$, *prove that*

$$(2n+1)\,\phi(n) = 2x^n\sqrt{(x-1)} + 2n\phi(n-1)$$

Solution:

We have

$$\phi(n) = \int_0^x \frac{x^n}{\sqrt{(x-1)}}dx = \int_0^x x^n(x-1)^{-1/2}dx$$

$$= \left[x^n \frac{(x-1)^{1/2}}{\frac{1}{2}} \right]_0^x - \int_0^x \frac{nx^{n-1}(x-1)^{1/2}}{\frac{1}{2}} dx,$$

integrating by parts taking $(x-1)^{-1/2}$ as the 2nd function

$$= 2x^n \sqrt{(x-1)} - 2n \int_0^x \frac{x^{n-1}(x-)}{\sqrt{(x-1)}} dx$$

$$= 2x^n \sqrt{(x-1)} - 2n \int_0^x \frac{x^n}{\sqrt{(x-1)}} dx + 2x \int_0^x \frac{x^{n-1}}{\sqrt{(x-1)}} dx$$

$$= 2x^n \sqrt{(x-1)} - 2n\phi(n) + 2n\phi(n-1).$$

Transposing the middle term to the left, we get

$(2n + 1)\,\phi(n) = 2x^n (x - 1)^{1/2} + 2n\phi\,(n - 1).$

4

DEFINITE INTEGRALS

4.1. DEFINITION

Sometimes in geometrical and other applications of integral calculus it becomes necessary to find the difference in the values of an integral of a function f (x) for two given values of the variable x, say, a and b. This difference is called the *definite integral* of f (x) from a to b or between the *limits* a and b.

This definite integral is denoted by

$$\int_a^b f(x)\,dx$$

and is read as "*the integral of f(x) with respect to x betwen the limits a and b*".

It is often written thus:

$$\int_a^b f(x)\,dx = [F(x)]_a^b = F(b) - F(a),$$

where F(x) is an integral of f(x), F(b) is the value of F(x) at x = b, and F(a) is the value of F(x) at x = a.

The number a is called the *lower limit* and the number b, the *upper limit* of integration. The interval (a, b) is called the *range of integration.*

4.2 FUNDAMENTAL PROPERTIES OF DEFINITE INTEGRALS

Property 1. *We have* $\int_a^b f(x)\,dx = \int_a^b f(t)\,dt$ *i.e., the value of a definite integral does not change with the change of variable of integration* (also called '**argument**') *provided the limits of integration remain the same.*

Proof:

Let $\int f(x)\,dx = F(x)$; then $\int f(t)\,dt = F(t)$

Now $\int_a^b f(x)\,dx = [F(x)]_a^b = F(b) - F(a)$ ...(1)

and $\int_a^b f(t)\,dt = [F(t)]_a^b = F(b) - F(a)$ ...(2)

From (1) and (2), we see that $\int_a^b f(x)\,dx = \int_a^b f(t)\,dt$.

Property 2. *We have* $\int_a^b f(x)\,dx = -\int_b^a f(x)\,dx$ *i.e., interchanging the limits of a definite integral does not change the absolute value but changes only the sign of the integral.*

Proof:

Let $\int f(x)\,dx = F(x)$. Then

$$\int_a^b f(x)\,dx = [F(x)]_a^b = F(b) - F(a) \quad ...(1)$$

Also $-\int_b^a f(x)\,dx = -[F(x)]_b^a = -[F(a) - F(b)] = F(b) - F(a)$ ...(2)

From (1) and (2), we see that $\int_a^b f(x)\,dx = -\int_b^a f(x)\,dx$.

Property 3. *We have* $\int_a^b f(x)\,dx = \int_a^c f(x)\,dx + \int_c^b f(x)\,dx$.

Proof:

Let $\int f(x)\,dx = F(x)$. The the R.H.S.

$= [F(x)]_a^c + [F(x)]_c^b = \{F(c) - F(a)\} + \{F(b) - F(c)\}$

$= F(b) - F(a) = \int_a^b f(x)\,dx = \text{L.H.S.}$

Notes:

1. This property also holds true even if the point c is exterior to the interval (a, b).
2. In place of one additional point c, we can take several points. Thus,

$$\int_a^b f(x)\,dx = \int_a^{c_1} f(x)\,dx + \int_{c_1}^{c_2} f(x)\,dx + \int_{c_2}^{c_3} f(x)\,dx$$
$$+ \dots + \int_{c_{r-1}}^{c_r} f(x)\,dx + \dots + \int_{c_n}^{b} f(x)\,dx.$$

Property 4. *We have* $\int_0^a f(x)\,dx = \int_0^a f(a-x)\,dx$

Proof:

Let $I = \int_0^a f(x)\,dx$

Put $x = a - t$, so that $dx = -dt$.

When $x = 0$, $t = a$ and when $x = a$, $t = 0$.

$\therefore\ I = \int_a^0 f(a-t)(-dt) = \int_0^a f(a-t)\,dt,$ [by property 2]

$= \int_0^a f(a-x)\,dx$ [by property 1]

Property 5. $\int_{-a}^{a} f(x)\,dx = 0$ *or* $\int_0^a f(x)\,dx$, *according as f(x) is an odd or an even function of x.*

Proof:

Odd and Even Functions : A function f(x) is said to be

(i) an odd function of x if $f(-x) = -f(x)$,

(ii) an even function of x if $f(-x) = f(x)$.

Now $\int_{-a}^{a} f(x)\,dx = \int_{-a}^{0} f(x)\,dx + \int_0^a f(x)\,dx$, by property 3. ...(1)

Let $u = \int_{-a}^{0} f(x)\,dx$. In the integral u,

put $x = -t$

so that $dx = -dt$.

Also $t = a$, when $x = -a$

and $t = 0$ when $x = 0$.

$\therefore\quad u = \int_a^0 f(-t)(-dt) = \int_0^a f(-t)\,dt$ [by property 2]

$= \int_0^a f(-x)\,dx,$ [by property 1]

$= -\int_0^a f(x)\,dx$ if f(x) is an odd function of x,

or $= \int_0^a f(x)\,dx$, if f(x) is an even function of x.

$\therefore$ from (1), we get

$\int_{-a}^{a} f(x)\,dx = -\int_0^a f(x)\,dx + \int_0^a f(x)\,dx = 0,$

if f(x) is an odd functon of x

and $\int_{-a}^{a} f(x)\,dx = \int_0^a f(x)\,dx + \int_0^a f(x)\,dx = 2\int_0^a f(x)\,dx,$

if f (x) is an even function of x.

Property 6. $\int_0^{2a} f(x)\,dx = 2\int_0^a f(x)\,dx, \text{ if } f(2a-x) = f(x)$

and $\int_0^{2a} f(x)dx = 0, \text{ if } f(2a-x) = -f(x).$

Proof :

We have $\int_0^{2a} f(x)\,dx = \int_0^a f(x)\,dx + \int_a^{2a} f(x)\,dx$

$= \int_0^a f(x)\,dx - \int_a^0 f(2a-y)\,dy,$ [putting x = 2a – y in the second integral and changing the limits]

$= \int_0^a f(x)\,dx + \int_0^a f(2a-y)\,dy,$

interchanging the limits in the second integral

$= \int_0^a f(x)\,dx + \int_0^a f(2a-x)\,dx,$

changing the argument from y to x in the second integral

$= 2\int_0^a f(x)\,dx, \text{ if } f(2a-x) = f(x)$ or $= 0, \text{ if } f(2a-x) = -f(x).$

Cor. $\int_0^{2a} f(x)\,dx = \int_0^a f(x)\,dx + \int_0^a f(2a-x)\,dx$

Remember:

(i) $\int_{-\pi/2}^{\pi/2} f(\sin x)\,dx = 2\int_0^{\pi/2} f(\sin x)\,dx$ or $= 0$

as if, f (sin x) is a *even* or an *odd* function respectively.

(ii) $\int_0^{\pi} f(\sin x)dx = 2\int_0^{\pi/2} f(\sin x)dx,$

[by property 6, because sin (π – x) = sin x]

(iii) $\int_{-\pi/2}^{\pi/2} f(\cos x)dx = 2\int_0^{\pi/2} f(\cos x)dx,$ [by property 5]

(iv) $\int_0^{\pi} f(\cos x)dx = 2\int_0^{\pi/2} f(\cos x)dx$ ... = 0,

as if, f (cos x) is an *even* or an *odd* function respectively.

(v) $\int_0^{\pi/2} f(\sin x)\,dx = \int_0^{\pi/2} f\left\{\sin\left(\frac{1}{2}\pi - x\right)\right\}dx$, [by property 4]

$$= \int_0^{\pi/2} f(\cos x)\,dx.$$

(vi) $\int_0^{\pi} \sin^m x \cos^n x\,dx = 2\int_0^{\pi/2} \sin^m x \cos^n x\,dx$ or $= 0$,

according as n is an even or an odd integer, (by property 6).

4.3. THE DEFINITE INTEGRAL AS THE LIMIT OF A SUM

So far integration has been defined as the inverse process of differentiation. But it is also possible to regard a definite integral as the limit of the sum of certain number of terms, when the number of terms tends to infinity and each term tends to zero.

Definition: *Let f(x) be a single valued continous function defined in the interval (a, b) where b > a and let the interval (a, b) divided into n equal parts each of h, so that nh = b – a; then we define*

$$\int_a^b \mathbf{f(x)\,dx = \lim h[f(a) + f(a+h) + f(a+2h) + \ldots}$$

$$\mathbf{+ f\{a + (n-1)h\}],}$$

when n → ∞, h → 0 and nh → b – a.

Thus, $\int_a^b f(x)\,dx = \lim_{h \to 0} h \sum_{r=0}^{n-1} f(a + rh)$, where $n \to \infty$ as $h \to 0$ and nh remains equal to b – a. We call $\int_a^b f(x)\,dx$ as the definite integral of f(x) w.r.t. x between the limts a and b.

4.4. SUMMATION OF SERIES WITH THE HELP OF DEFINITE INTEGRALS

The definition of a definite integral as the limit of a sum (7.3) helps us to evaluate the limit of the sums of some special types of series. We know that

$$\int_a^b f(x)\,dx = \lim_{n \to \infty} h[f(a) + f(a+h) + \ldots + f\{a + (n-1)h\}]$$

$$= \lim_{n \to \infty} h \sum_{r=0}^{n-1} f(a + rh), \text{ where } nh = b - a.$$

Putting a = 0 and b = 1,

so that $h = \left(\frac{1}{n}\right)$, we get

$$\int_0^1 f(x)dx = \lim_{n\to\infty} \frac{1}{n}\sum_{r=0}^{n-1} f\left(\frac{r}{n}\right).$$

Thus, the limit of the sum of a series can be expressed in the form of a definite integral provided the series has the following properties:

(a) Each term of the series should have $\left(\frac{1}{n}\right)$ as a common factor which tends to zero as $n \to \infty$.

(b) The general term of the series should be the product of $1/n$ and a function $f(r/n)$ of r/n, so that the various terms of the series can be obtained from it by giving different values to r, say $r = 0, 1, 2..., n - 1$.

(c) There should be n terms in the series, but if however the number of terms differs by a finite number from n, then the required limit does not change because each term tends to zero. Thus

$$\lim_{n\to\infty} \frac{1}{n}\sum_{r=p}^{n+q} f\left(\frac{r}{n}\right) = \int_0^1 f(x)dx,$$

if p and q are independent of n.

Working Rule:

(i) Write down the general term [say rth term or (r – 1)th term etc. as convenient] of the series. Take out $\left(\frac{1}{n}\right)$ as a factor from the general term and thus write the series in the form $\frac{1}{n}\sum_{r=0}^{n-1} f\left(\frac{r}{n}\right)$. We may have some other limits of r in the summation; for example, r may vary from 1 to n or from 0 to 2n, etc.

(ii) Now the evaluate $\lim_{n\to\infty} \frac{1}{n}\sum_{r=0}^{n-1} f\left(\frac{r}{n}\right)$, replace $\frac{r}{n}$ by x, $\frac{1}{n}$ by dx and $\lim_{n\to\infty} \Sigma$ by the sign of integration *i.e.*, by $\int$.

(iii) To find the limits of integration of x first note carefully the limits of r in the summation $\Sigma f(r/n)$. Divide these limits by n to get the values of r/n. Take limits of these values of r/n as $n \to \infty$ and get the limits of integration of x.

MISCELLANEOUS EXAMPLES

Example 1:

Evaluate

$$\lim_{n\to\infty}\left[\frac{n^{1/2}}{n^{n/2}}+\frac{n^{1/2}}{(n+3)^{3/2}}+\frac{n^{1/2}}{(n+6)^{3/2}}+\ldots+\frac{n^{1/2}}{\{n+3(n-1)\}^{3/2}}\right].$$

Solution:

Here the general term $=\dfrac{n^{1/2}}{(n+3r)^{3/2}}$

$$=\frac{1}{n\left\{1+\left(\frac{3r}{n}\right)\right\}^{3/2}},\qquad \text{(and r varies from 0 to n − 1)}$$

$\therefore$ the given limit $=\lim\limits_{n\to\infty}\sum\limits_{r=0}^{n}\dfrac{1}{n}\cdot\dfrac{1}{\left\{1+\left(\frac{3r}{n}\right)\right\}^{3/2}}$

$$=\int_0^1\frac{dx}{(1+3x)^{3/2}}=-\frac{2}{3}\cdot\left[\frac{1}{(1+3x)^{1/2}}\right]_0^1$$

$$=-\frac{2}{3}\left[\frac{1}{2}-1\right]=\frac{1}{3}.$$

Example 2:

Evaluate

$$\lim_{n\to\infty}\left[\frac{1}{n}+\frac{1}{\sqrt{(n^2-1^2)}}+\frac{1}{\sqrt{(n^2-2^2)}}+\ldots+\frac{1}{\sqrt{\{n^2-(n-1)^2\}}}\right].$$

Solution:

Here the general term

$$=\frac{1}{\sqrt{(n^2-r^2)}}=\frac{1}{n}\cdot\frac{1}{\sqrt{\left\{1-\left(\frac{r}{n}\right)^2\right\}}},\ \text{and r varies from 0 to (n − 1).}$$

$\therefore$ the given limit $=\lim\limits_{n\to\infty}\sum\limits_{r=0}^{n-1}\dfrac{1}{n}\cdot\dfrac{1}{\sqrt{\left\{1-\left(\frac{r}{n}\right)^2\right\}}}$

$$= \int_0^1 \frac{1}{\sqrt{(1-x^2)}} dx = \left[\sin^{-1} x\right]_0^1$$

$$= \sin^{-1} 1 - \sin^{-1} 0 = \frac{1}{2}\pi.$$

Example 3:

Evaluate

$$\lim_{n\to\infty}\left[\frac{1}{\sqrt{(n^2-1^2)}} + \frac{1}{\sqrt{(n^2-2^2)}} + \ldots + \frac{1}{\sqrt{\{n^2-(n-1)^2\}}}\right].$$

Solution:

Do your self.

Example 4:

Evaluate

$$\lim_{n\to\infty}\left[\frac{1}{n^2}\sec^2\frac{1}{n^2} + \frac{2}{n^2}\sec^2\frac{4}{n^2} + \frac{3}{n^2}\sec^2\frac{9}{n^2} + \ldots + \frac{1}{n}\sec^2 1\right].$$

Solution:

Here the rth term

$$= \frac{r}{n^2}\sec^2\frac{r^2}{n^2} = \frac{1}{n}\cdot\left\{\frac{r}{n}\sec^2\frac{r^2}{n^2}\right\},$$ and r varies from 1 to n.

$\therefore$ the given limit $= \lim_{n\to\infty}\sum_{r=1}^{n}\frac{1}{n}\cdot\left\{\frac{r}{n}\sec^2\frac{r^2}{n^2}\right\}$

$= \int_0^1 x\sec^2 x^2 dx = \frac{1}{2}\int_0^1 \sec^2 t\, dt$, putting $x^2 = t$ so that

$2x\,dx = dt$ and the limits for t are 0 to 1

$$= \frac{1}{2}[\tan t]_0^1 = \frac{1}{2}(\tan 1 - \tan 0) = \frac{1}{2}\tan 1.$$

Example 5:

Evaluate

$$\lim_{n\to\infty}\frac{1}{n}\left[\sin^{2k}\frac{\pi}{2n} + \sin^{2k}\frac{2\pi}{2n} + \sin^{2k}\frac{3\pi}{2n} + \ldots + \sin^{2k}\frac{\pi}{2}\right].$$

Solution:

Here the rth term $= \frac{1}{n}\cdot\sin^{2k}\frac{r\pi}{2n}$, and r varies from 1 to n.

$\therefore$ the given limit $= \lim\limits_{n\to\infty} \sum\limits_{r=1}^{n} \frac{1}{n}\cdot \sin^{2k} \frac{r\pi}{2n}$

$= \int_0^1 \sin^{2k}\left(\frac{\pi}{2}\cdot x\right)dx = \frac{2}{\pi}\int_0^{\pi/2} \sin^{2k} t\,dt$

so that $\frac{1}{2}\pi\,dx = dt$, and the limits for t are 0 to $\frac{\pi}{2}$

$= \frac{2}{\pi}\cdot\frac{(2k-1)}{2k}\cdot\frac{(2k-3)}{(2k-2)}\cdots\frac{3}{4}\cdot\frac{1}{2}\cdot\frac{\pi}{2}$, by Walli's formula

$= \frac{(2k-1)(2k-3)\ldots 3.1}{2k.(2k-2)\ldots 4.2}$.

Example 6:

Evaluate $\lim\limits_{n\to\infty} \sum\limits_{r=1}^{n} \frac{r^3}{r^4+n^4}$.

Solution:

Here $\lim\limits_{n\to\infty} \sum\limits_{r=1}^{n} \frac{r^3}{r^4+n^4} = \lim\limits_{n\to\infty} \sum\limits_{r=1}^{n} \frac{1}{n^4}\left\{\frac{r^3}{\left(\frac{r}{n}\right)^4+1}\right\}$

$= \lim\limits_{n\to\infty} \sum\limits_{r=1}^{n} \frac{1}{n}\cdot\left\{\frac{\left(\frac{r}{n}\right)^3}{\left(\frac{r}{n}\right)^4+1}\right\}$ **(Note)**

$= \int_0^1 \frac{x^3}{x^4+1}dx = \frac{1}{4}\left[\log\left(1+x^4\right)\right]_0^1 = \frac{1}{4}\log 2$.

Example 7:

Prove that $\lim\limits_{n\to\infty} \sum\limits_{r=0}^{n-1} \frac{n}{r^2+n^2} = \frac{\pi}{4}$.

Solution:

Do your self.

Example 8:

Evaluate $\lim\limits_{n\to\infty} \sum\limits_{r=0}^{n-1} \frac{1}{n}\cdot\sqrt{\left\{\frac{1+\left(\frac{r}{n}\right)}{1-\left(\frac{r}{n}\right)}\right\}}$.

Solution:

Here $\lim\limits_{n\to\infty}\sum\limits_{r=1}^{n-1}\frac{1}{n}\sqrt{\left(\frac{n+r}{n-r}\right)}$

$$=\lim_{n\to\infty}\sum_{r=1}^{n-1}\frac{1}{n}\cdot\sqrt{\left\{\frac{1+\left(\frac{1}{n}\right)}{1-\left(\frac{r}{n}\right)}\right\}}$$

$$=\int_0^1\sqrt{\left(\frac{1+x}{1-x}\right)}dx=\int_0^1\frac{(1+x)}{\sqrt{(1-x^2)}}dx. \qquad \textbf{(Note)}$$

Now put $x = \sin\theta$

so that $dx = \cos\theta\, d\theta$.

Also $\theta = 0$ when $x = 0$

and $\theta=\frac{\pi}{2}$ when $x = 1$.

$\therefore$ the required limit $=\int_0^{\pi/2}\frac{1+\sin\theta}{\cos\theta}\cos\theta\, d\theta$

$$=\int_0^{\pi/2}(1+\sin\theta)d\theta=[\theta-\cos\theta]_0^{\pi/2}$$

$$=\left(\frac{\pi}{2}-0\right)-(0-1)=\frac{\pi}{2}+1.$$

Example 9:

Prove that

$$\lim_{n\to\infty}\left[\frac{1}{\sqrt{(2n-1^2)}}+\frac{1}{\sqrt{(4n-2^2)}}+\ldots+\frac{1}{n}\right]=\frac{\pi}{2}.$$

Solution:

Here the rth term

$$=\frac{1}{\sqrt{(2nr-r^2)}}=\frac{1}{n}\cdot\frac{1}{\sqrt{\left\{2\left(\frac{r}{n}\right)-\left(\frac{r}{n}\right)^2\right\}}}, \text{ and r varies from 1 to n.}$$

$\therefore$ the given limit

$$=\lim_{n\to\infty}\sum_{r=1}^{n}\frac{1}{n}\cdot\frac{1}{\sqrt{\left\{2\left(\frac{1}{n}\right)-\left(\frac{r}{n}\right)^2\right\}}}$$

$$= \int_0^1 \frac{dx}{\sqrt{(2x-x^2)}} = \int_0^1 \frac{dx}{\sqrt{\{1-(x-1)^2\}}} [\sin^{-1}(x-1)]_0^1$$

$$= \sin^{-1} 0 - \sin^{-1}(-1) = 0 + \sin^{-1} 1 = \frac{\pi}{2}.$$

Example 10:

Evaluate $\lim_{n\to\infty}\left[\tan\frac{\pi}{2n}\tan\frac{2\pi}{2n}\tan\frac{3\pi}{2n}...\tan\frac{n\pi}{2n}\right]^{1/n}$

Solution:

Let $P = \lim_{n\to\infty}\left[\tan\frac{\pi}{2n}\tan\frac{2\pi}{2n}...\tan\frac{n\pi}{2n}\right]^{1/n}$.

Then

$$\log P = \lim_{n\to\infty}\frac{1}{n}\left[\log\left(\tan\frac{\pi}{2n}\right)+\log\left(\tan\frac{2\pi}{2n}\right)+...+\log\left(\tan\frac{n\pi}{2n}\right)\right]$$

$$= \lim_{n\to\infty}\frac{1}{n}\sum_{r=1}^{n}\log\left(\tan\frac{r\pi}{2n}\right)$$

$$= \int_0^1 \log\left\{\tan\frac{\pi}{2}x\right\}dx = \frac{2}{\pi}\int_0^{\pi/2}\log(\tan\theta)d\theta$$

putting $\left(\frac{\pi x}{2}\right) = \theta$ and changing the limits accordingly

$$= \frac{2}{\pi}\int_0^{\pi/2}\log\left(\frac{\sin\theta}{\cos\theta}\right)d\theta$$

$$= \frac{2}{\pi}\int_0^{\pi/2}\log\sin\theta\,d\theta - \frac{2}{\pi}\int_0^{\pi/2}\log\cos\theta\,d\theta$$

$$= \frac{2}{\pi}\int_0^{\pi/2}\log\sin\left(\frac{\pi}{2}-\theta\right)d\theta - \frac{2}{\pi}\int_0^{\pi/2}\log\cos\theta\,d\theta,$$

$$\left[\because \int_0^a f(x)dx = \int_0^a f(a-x)dx\right]$$

$$= \frac{2}{\pi}\int_0^{\pi/2}\log\cos\theta\,d\theta - \frac{2}{\pi}\int_0^{\pi/2}\log\cos\theta\,d\theta = 0.$$

$\therefore$ $P = e^0 = 1$.

Example 11:

Evaluate the limit

$$\lim_{n\to\infty}\left(1+\frac{1}{n^2}\right)^{2/n^2}\left(1+\frac{2^2}{n^2}\right)^{4/n^2}\left(1+\frac{3^2}{n^2}\right)^{6/n^2}...\left(1+\frac{n^2}{n^2}\right)^{2n/n^2}.$$

Solution:

Let the required limit be P. Then

$$\log P = \lim_{n\to\infty}\left[\frac{2}{n^2}\log\left(1+\frac{1}{n^2}\right)+\frac{4}{n^2}\log\left(1+\frac{2^2}{n^2}\right)+\ldots+\frac{2n}{n^2}\log\left(1+\frac{n^2}{n^2}\right)\right]$$

$$= \lim_{n\to\infty}\sum_{r=1}^{n}\frac{2r}{n^2}\log\left(1+\frac{r^2}{n^2}\right)$$

$$= \lim_{n\to\infty}\frac{1}{n}\sum_{r=1}^{n}2\left(\frac{r}{n}\right)\log\left\{1+\left(\frac{r}{n}\right)^2\right\} = \int_0^1 2x\log(1+x^2)\,dx.$$

Now put $\quad 1 + x^2 = t,$

so that $\quad 2x\,dx = dt.$

When $\quad x = 0, t = 1$

and when $\quad x = 1, t = 2.$

$$\therefore\ \log P = \int_1^2 \log t\,dt$$

$$= [t\log t]_1^2 - \int_1^2 t\cdot\frac{1}{t}dt,$$

integrating by parts taking 1 as the second function

$$= (2\log 2 - \log 1) - \int_1^2 dt = 2\log 2 - [t]_1^2$$

$$= \log 2^2 - (2-1) = \log 4 - 1 = \log 4 - \log e = \log\left(\frac{4}{e}\right).$$

$$\therefore\ P = \frac{4}{e}.$$

Example 12:

Evaluate $\displaystyle\int_0^{\pi}\frac{x\,dx}{a^2\cos^2 x + b^2\sin^2 x}$.

Solution:

$$\text{Let } I = \int_0^{\pi}\frac{x\,dx}{a^2\cos^2 x + b^2\sin^2 x}. \qquad \ldots(1)$$

$$\text{Then } I = \int_0^{\pi}\frac{(\pi - x)\,dx}{a^2\cos^2(\pi - x) + b^2\sin^2(\pi - x)},$$

$$\left[\because \int_0^a f(x)\,dx = \int_0^a f(a-x)\,dx\right]$$

$$= \int_0^{\pi}\frac{(\pi - x)\,dx}{a^2\cos^2 x + b^2\sin^2 x}. \qquad \ldots(2)$$

Adding (1) and (2), we get

$$2I = \int_0^{\pi} \frac{x + (\pi - x)}{a^2\cos^2 x + b^2 \sin^2 x} dx$$

$$= \pi \int_0^{\pi} \frac{dx}{a^2\cos^2 x + b^2 \sin^2 x}$$

$$= 2\pi \int_0^{\pi/2} \frac{dx}{a^2\cos^2 x + b^2 \sin^2 x},$$ by a property of definite integrals. (Refer prop. 6)

$$\therefore\ I = \pi \int_0^{\pi/2} \frac{dx}{a^2\cos^2 x + b^2 \sin^2 x}.$$

$$= \pi \int_0^{\pi/2} \frac{\sec^2 x\, dx}{a^2 + b^2 \tan^2 x},$$ dividing the numerator and the denominator by $\cos^2 x$.

Now put b tan x = t.

Then b $\sec^2$ x dx = dt.

Also when x = 0, t = 0

and when $x \to \frac{\pi}{2}$, $t \to \infty$.

$$\therefore\ I = \frac{\pi}{b} \int_0^{\infty} \frac{dt}{a^2 + t^2} = \frac{\pi}{b} \cdot \frac{1}{a} \left[\tan^{-1} \frac{t}{a}\right]_0^{\infty}$$

$$= \frac{\pi}{ab} \left[\tan^{-1} \infty - \tan^{-1} 0\right]$$

$$= \frac{\pi}{ab} \left[\frac{\pi}{2} - 0\right] = \frac{\pi^2}{2ab}.$$

Example 13:

Show that $\int_0^{\pi} \frac{x\,dx}{\left(a^2\cos^2 x + b^2 \sin^2 x\right)^2} = \frac{\pi^2\left(a^2 + b^2\right)}{4a^3b^3}$

Solution:

Let $I = \int_0^{\pi} \frac{x\,dx}{\left[a^2\cos^2 x + b^2 \sin^2 x\right]^2}$

$$= \int_0^{\pi} \frac{(\pi - x)dx}{\left[a^2\cos^2 (\pi - x) + b^2 \sin^2 (\pi - x)\right]^2},$$ (Refer prop. 4)

$$=\int_0^{\pi}\frac{(\pi-x)dx}{\left[a^2\cos^2x+b^2\sin^2x\right]^2}=\int_0^{\pi}\frac{\pi\, dx}{\left[a^2\cos^2x+b^2\sin^2x\right]^2}-I$$

$$\therefore\ 2I=\int_0^{\pi}\frac{\pi\, dx}{\left[a^2\cos^2x+b^2\sin^2x\right]^2}$$

$$=2\int_0^{\pi/2}\frac{\pi\, dx}{\left[a^2\cos^2x+b^2\sin^2x\right]^2},\qquad \text{(Refer prop. 6)}$$

or $I=\pi\int_0^{\pi/2}\frac{\sec^4x\, dx}{\left(a^2+b^2\tan^2x\right)^2}$, dividing Nr. and the Dr. by $\cos^4 x$

$$=\pi\int_0^{\pi/2}\frac{\left(1+\tan^2x\right)\sec^2x}{\left(a^2+b^2\tan^2x\right)^2}dx$$

Now put b tan x = a tan θ,
so that $b\sec^2 x\, dx = a\sec^2\theta\, d\theta$.
Also when x = 0, θ = 0

and when $x=\frac{1}{2}\pi,\ \theta=\frac{1}{2}\pi$.

$$\therefore\ I=\pi\int_0^{\pi/2}\frac{\left\{1+\left(a^2/b^2\right)\tan^2\theta\right\}.(a/b)\sec^2\theta\, d\theta}{a^4\sec^4\theta}$$

$$=\frac{\pi}{a^3b^3}\int_0^{\pi/2}\left(b^2\cos^2\theta+a^2\sin^2\theta\right)d\theta$$

$$=\frac{\pi}{a^3b^3}\left[b^2\cdot\frac{1}{2}\cdot\frac{\pi}{2}+a^2\cdot\frac{1}{2}\cdot\frac{\pi}{2}\right],\qquad \text{(by Walli's formula)}$$

$$=\frac{\pi^2}{4a^3b^3}\left(a^2+b^2\right).$$

Example 14:

Show that $\int_0^{\pi}\frac{x\, dx}{a^2-\cos^2x}=\frac{\pi^2}{2a\sqrt{\left(a^2-1\right)}}$, *(a > 1).*

Solution:

Let $I=\int_0^{\pi}\frac{x\, dx}{a^2-\cos^2x}=\int_0^{\pi}\frac{(\pi-x)dx}{a^2-\cos^2(\pi-x)}$, (Refer prop. 4)

$$= \pi \int_0^{\pi} \frac{dx}{a^2 - \cos^2 x} - I.$$

$$\therefore \; 2I = \int_0^{\pi} \frac{\pi\, dx}{a^2 - \cos^2 x} = 2\int_0^{\pi/2} \frac{\pi\, dx}{a^2 - \cos^2 x}, \qquad \text{(Refer prop. 6)}$$

or $I = \pi \int_0^{\pi/2} \frac{\sec^2 x\, dx}{a^2 \sec^2 x - 1}$, dividing the Nr. and Dr. by $\cos^2 x$

$$= \pi \int_0^{\pi/2} \frac{\sec^2 x\, dx}{a^2(1 + \tan^2 x) - 1} = \pi \int_0^{\pi/2} \frac{\sec^2 x\, dx}{(a^2 - 1) + a^2 \tan^2 x}.$$

Now put $a \tan x = t$,
so that $a \sec^2 x \, dx = dt$.
Also $t = 0$ when $x = 0$
and $t \to \infty$ when $x \to \frac{1}{2}\pi$.

$$\therefore \; I = \frac{\pi}{a} \int_0^{\infty} \frac{dt}{(a^2 - 1) + t^2} = \frac{\pi}{a} \cdot \frac{1}{\sqrt{(a^2 - 1)}} \left[\tan^{-1} \left\{ \frac{t}{\sqrt{(a^2 - 1)}} \right\} \right]_0^{\infty}$$

$$= \frac{\pi}{a\sqrt{(a^2 - 1)}} \left[\tan^{-1} \infty - \tan^{-1} 0 \right]$$

$$= \frac{\pi}{a\sqrt{(a^2 - 1)}} \left[\frac{\pi}{2} - 0 \right] = \frac{\pi^2}{2a\sqrt{(a^2 - 1)}}.$$

Example 15:

Evaluate $\int_0^{\pi} \frac{x\, dx}{1 + \cos^2 x}$.

Solution:

Let $I = \int_0^{\pi} \frac{x\, dx}{1 + \cos^2 x} = \int_0^{\pi} \frac{(\pi - x)\, dx}{1 + \cos^2(\pi - x)}$, (Refer prop. 4)

$$= \int_0^{\pi} \frac{(\pi - x)\, dx}{1 + \cos^2 x} = \int_0^{\pi} \frac{\pi\, dx}{1 + \cos^2 x} - \int_0^{\pi} \frac{x\, dx}{1 + \cos^2 x}$$

$$= \pi \int_0^{\pi} \frac{dx}{1 + \cos^2 x} - I.$$

$$\therefore \; 2I = \int_0^{\pi} \frac{dx}{1 + \cos^2 x} = 2\pi \int_0^{\pi/2} \frac{dx}{1 + \cos^2 x}, \qquad \text{(Refer prop. 6)}$$

or $I = \pi \int_0^{\pi/2} \frac{\sec^2 x\, dx}{\sec^2 x + 1}$, dividing the Nr. and Dr. by $\cos^2 x$

$$= \pi \int_0^{\pi/2} \frac{\sec^2 x\, dx}{2 + \tan^2 x}.$$

Now put $\tan x = t$,

so that $\sec^2 x\ dx = dt$.

Also $t = 0$ when $x = 0$

and $t \to \infty$ when $x \to \frac{1}{2}\pi$.

$$\therefore\ I = \pi \int_0^{\infty} \frac{dt}{2 + t^2} = \pi \cdot \frac{1}{\sqrt{2}} \left[\tan^{-1} \frac{t}{\sqrt{2}} \right]_0^{\infty}$$

$$= \frac{\pi}{\sqrt{2}} \left[\tan^{-1} \infty - \tan^{-1} 0 \right]$$

$$= \frac{\pi}{\sqrt{2}} \left[\frac{\pi}{2} - 0 \right] = \frac{\pi^2}{2\sqrt{2}} = \frac{\pi^2 \sqrt{2}}{4}.$$

Example 16:

Evaluate $\int_0^{\pi/2} \frac{\cos x - \sin x}{1 + \sin x \cos x} dx$.

Solution:

Let $I = \int_0^{\pi/2} \frac{\cos x - \sin x}{1 + \sin x \cos x} dx$.

Then $I = \int_0^{\pi/2} \frac{\cos\left(\frac{1}{2}\pi - x\right) - \sin\left(\frac{1}{2}\pi - x\right)}{1 + \sin\left(\frac{1}{2}\pi - x\right) \cos\left(\frac{1}{2}\pi - x\right)} dx$, [Refer prop. 4]

$$= \int_0^{\pi/2} \frac{\sin x - \cos x}{1 + \cos x \sin x} dx = -\int_0^{\pi/2} \frac{\cos x - \sin x}{1 + \sin x \cos x} dx = -I.$$

$\therefore\ 2I = 0$ or $I = 0$.

Example 17:

Evaluate $\int_0^{\pi/2} (\sin x - \cos x) \log (\sin x + \cos x) dx$.

Solution:

Let $I = \int_0^{\pi/2} (\sin x - \cos x) \log (\sin x + \cos x) dx$.

Then $I = \int_0^{\pi/2} (\cos x - \sin x) \log (\sin x + \cos x) dx$

$$\left[\because \int_0^a f(x)\,dx = \int_0^a f(a-x)\,dx \text{ Note that } \sin\left(\frac{1}{2}\pi - x\right) = \cos x\right.$$

$$\left.\text{and } \cos\left(\frac{1}{2}\pi - x\right) = \sin x\right]$$

$$= -\int_0^{\pi/2} (\sin x - \cos x)\log(\sin x + \cos x)\,dx = -I.$$

$\therefore\ 2I = 0 \quad \text{or} \quad I = 0.$

Example 18:

Evaluate $\lim\limits_{n\to\infty} \sum\limits_{r=1}^{n} \dfrac{\sqrt{n}}{\sqrt{r}.\left(3\sqrt{r} + 4\sqrt{n}\right)^2}.$

Solution:

The given limit

$$= \lim_{n\to\infty} \sum_{r=1}^{n} \frac{1}{n}\cdot\frac{1}{\sqrt{\left(\frac{r}{n}\right)}.\left\{3\sqrt{\left(\frac{r}{n}\right)} + 4\right\}^2}$$

$= \displaystyle\int_0^1 \frac{dx}{\sqrt{x}.\left(3\sqrt{x}+4\right)^2}$. Now put $3\sqrt{x} + 4 = t$,

so that $3.\left(\frac{1}{2}\sqrt{x}\right)dx = dt$.

The limits for t are 4 to 7.

$\therefore$ the required limit

$$= \frac{2}{3}\int_4^7 \frac{dt}{t^2} = -\frac{2}{3}\left[\frac{1}{t}\right]_4^7 = -\frac{2}{3}\left[\frac{1}{7} - \frac{1}{4}\right] = \frac{1}{14}.$$

Example 19:

Evaluate $\lim\limits_{n\to\infty} \sum\limits_{r=1}^{n} \dfrac{n^2}{\left(n^2 + r^2\right)^{3/2}}.$

Solution:

The given limit

$$= \lim_{n\to\infty} \sum_{r=1}^{n} \frac{1}{n}\cdot\frac{1}{\left[1 + \left(\frac{r}{n}\right)^2\right]^{3/2}}$$

$= \displaystyle\int_0^1 \frac{dx}{\left(1+x^2\right)^{3/2}} = \int_0^{\pi/4} \frac{\sec^2\theta\, d\theta}{\sec^3\theta}$, putting $x = \tan\theta$

so that $dx = \sec^2\theta\, d\theta$

and the limits for θ are 0 to $\frac{\pi}{4}$

$$= \int_0^{\pi/4} \cos\theta\, d\theta = [\sin\theta]_0^{\pi/4} = \sin\frac{\pi}{4} - \sin 0 = \frac{1}{\sqrt{2}}.$$

Example 20:

Evaluate $\int_0^{\pi/2} \sin 2x \log \tan x\, dx.$

Solution:

Let $I = \int_0^{\pi/2} \sin 2x \log \tan x\, dx.$ ...(1)

Then $I = \int_0^{\pi/2} \sin 2\left(\frac{1}{2}\pi - x\right) \log \tan\left(\frac{1}{2}\pi - x\right),$ (Refer prop. 4)

$$= \int_0^{\pi/2} \sin(\pi - x) \log \cot x\, dx$$

$$= \int_0^{\pi/2} \sin 2x \log \cot x\, dx. \quad ...(2)$$

Adding (1) and (2), we get

$$2I = \int_0^{\pi/2} \sin 2x(\log \tan x + \log \cot x)\, dx$$

$$= \int_0^{\pi/2} \sin 2x \log(\tan x \cot x)\, dx$$

$$= \int_0^{\pi/2} (\sin 2x).\log 1\, dx$$

$$= \int_0^{\pi/2} 0.\sin 2x\, dx$$

$$= 0 \times \int_0^{\pi/2} \sin 2x\, dx = 0.$$

$\therefore$ I = 0.

Example 21:

Show that $\int_0^{\pi/2} \frac{\sin x - \cos x}{\sin x + \cos x} = 0.$

Solution:

Do your self.

Example 22:

Evaluate $\int_0^{\pi} \frac{x\, dx}{1 + \sin x}.$

Solution:

Let $I=\int_0^{\pi}\frac{x\,dx}{1+\sin x}=\int_0^{\pi}\frac{(\pi-x)\,dx}{1+\sin(\pi-x)}$, (Refer prop. 4)

$$=\int_0^{\pi}\frac{(\pi-x)}{1+\sin x}dx=\int_0^{\pi}\frac{\pi}{1+\sin x}dx-\int_0^{\pi}\frac{x}{1+\sin x}dx$$

$$=\pi\int_0^{\pi}\frac{1}{1+\sin x}dx-I.$$

$\therefore\ 2I=\pi\int_0^{\pi}\frac{dx}{1+\sin x}=2\pi\int_0^{\pi/2}\frac{dx}{1+\sin x}$, (Refer prop. 6)

or $I=\pi\int_0^{\pi/2}\frac{dx}{1+\sin x}=\pi\int_0^{\pi/2}\frac{dx}{1+\sin\left(\frac{1}{2}\pi-x\right)}$, (Refer prop. 4)

$$=\pi\int_0^{\pi/2}\frac{dx}{1+\cos x}=\pi\int_0^{\pi/2}\frac{dx}{2\cos^2\frac{1}{2}x}=\pi\int_0^{\pi/2}\frac{1}{2}\sec^2\frac{1}{2}x\,dx$$

$$=\pi\left[\tan\frac{1}{2}x\right]_0^{\pi/2}=\pi\left[\tan\frac{1}{2}\pi-\tan 0\right]=\pi(1-0)=\pi.$$

Example 23:

Prove that $\int_0^{\pi}\frac{x\sin x}{1+\sin x}dx=\pi\left(\frac{\pi}{2}-1\right)$.

Solution:

Let $I=\int_0^{\pi}\frac{x\sin x}{1+\sin x}dx=\int_0^{\pi}\frac{(\pi-x)\sin(\pi-x)}{1+\sin(\pi-x)}dx$, (Refer prop. 4)

$$=\int_0^{\pi}\frac{(\pi-x)\sin x}{1+\sin x}dx=\pi\int_0^{\pi}\frac{\sin x}{1+\sin x}dx-\int_0^{\pi}\frac{x\sin x}{1+\sin x}dx$$

$$=\pi\int_0^{\pi}\frac{\sin x}{1+\sin x}dx-I.$$

$\therefore\ 2I=\pi\int_0^{\pi}\frac{\sin x}{1+\sin x}dx=2\pi\int_0^{\pi/2}\frac{\sin x}{1+\sin x}dx$, (Refer prop. 6)

or $I=\pi\int_0^{\pi/2}\frac{(1+\sin x)-1}{1+\sin x}dx=\pi\int_0^{\pi/2}\left[1-\frac{1}{1+\sin x}\right]dx$

$$=\pi\int_0^{\pi/2}dx-\pi\int_0^{\pi/2}\frac{dx}{1+\sin x}$$

$$= \pi[x]_0^{\pi/2} - \pi\int_0^{\pi/2} \frac{dx}{1+\sin\left(\frac{1}{2}\pi - x\right)} \qquad \text{(Refer prop. 4)}$$

$$= \pi\left(\frac{\pi}{2} - 0\right) - \pi\int_0^{\pi/2} \frac{dx}{1+\cos x} = \frac{\pi^2}{2} - \pi\int_0^{\pi/2} \frac{dx}{2\cos^2\frac{1}{2}x}$$

$$= \frac{\pi^2}{2} - \pi\int_0^{\pi/2} \frac{1}{2}\sec^2\frac{1}{2}x\,dx = \frac{\pi^2}{2} - \pi\left[\tan\frac{1}{2}x\right]_0^{\pi/2}$$

$$= \frac{\pi^2}{2} - \pi\left[\tan\frac{1}{4}\pi - \tan 0\right] = \frac{\pi^2}{2} - \pi = \pi\left(\frac{\pi}{2} - 1\right).$$

Example 24:

Evaluate $\int_0^{\pi} \frac{x \sin x}{\left(1+co^2 x\right)} dx$

Solution:

Let $I = \int_0^{\pi} \frac{x \sin x}{1+\cos^2 x} dx.$ ...(1)

Then $I = \int_0^{\pi} \frac{(\pi - x)\sin(\pi - x)}{1+\cos^2(\pi - x)} dx, \quad \left[\because \int_0^a f(x)dx = \int_0^a f(a-x)dx\right]$

$$= \int_0^{\pi} \frac{(\pi - x)\sin x}{1+\cos^2 x} dx. \qquad ...(2)$$

Adding (1) and (2), we get

$$2I = \int_0^{\pi} \frac{\pi \sin x}{1+\cos^2 x} dx = \pi\int_0^{\pi} \frac{\sin x}{1+\cos^2 x} dx$$

$$= 2\pi\int_0^{\pi/2} \frac{\sin x}{1+\cos^2 x} dx, \qquad \text{(Refer prop. 6)}$$

or $I = \pi\int_0^{\pi/2} \frac{\sin x}{1+\cos^2 x} dx.$

Now put $\cos x = t$,

so that $-\sin x\, dx = dt$.

Also $t = 1$ when $x = 0$

and $t = 0$ when $x = \frac{1}{2}\pi$.

$$\therefore\ I = \pi\int_1^0 \frac{-dt}{1+t^2} = \pi\int_0^1 \frac{dt}{1+t^2} = \pi\left[\tan^{-1} t\right]_0^1$$

$$= \pi\left(\tan^{-1}1 - \tan^{-1}0\right) = \pi\left(\frac{1}{4}\pi - 0\right) = \frac{1}{4}\pi^2.$$

Example 25:

Show that $\int_0^{\pi} \frac{x\tan x}{\sec x + \cos x}dx = \frac{1}{4}\pi^2$.

Solution:

The given integral

$$I = \int_0^{\pi} \frac{x.\left(\frac{\sin x}{\cos x}\right)}{\left(\frac{1}{\cos x}\right) + \cos x}dx$$

$$= \int_0^{\pi} \frac{x\sin x}{1+\cos^2 x}dx = \frac{\pi^2}{4}.$$

Example 26:

Show that $\int_0^{\pi} \frac{x\tan x\,dx}{\sec x + \tan x} = \pi\left(\frac{1}{2}\pi - 1\right)$.

Solution:

Do your self.

$$\text{Let } I = \int_0^{\pi} \frac{x\tan x}{\sec x + \tan x}dx = \int_0^{\pi} \frac{x\left(\frac{\sin x}{\cos x}\right)}{\left(\frac{1}{\cos x}\right) + \left(\frac{\sin x}{\cos x}\right)}dx$$

$$= \int_0^{\pi} \frac{x\sin x}{1+\sin x}dx.$$

Example 27:

Evaluate $\int_0^{\pi} x\sin^6 x\cos^4 x\,dx$.

Solution:

$$\text{Let } I = \int_0^{\pi} x\sin^6 x\cos^4 x\,dx$$

$$= \int_0^{\pi} (\pi - x)\sin^6(\pi - x)\cos^4(\pi - x)dx, \qquad \text{(Refer prop. 4)}$$

$$= \int_0^{\pi} (\pi - x)\sin^6 x\cos^4 x\,dx$$

$$= \int_0^{\pi} \pi \sin^6 x \cos^4 x\, dx - \int_0^{\pi} x \sin^6 x \cos^4 x\, dx$$

$$= \pi \int_0^{\pi} \sin^6 x \cos^4 x\, dx - I.$$

$$\therefore\ 2I = \pi \int_0^{\pi} \sin^6 x \cos^4 x\, dx = 2\pi \int_0^{\pi/2} \sin^6 x \cos^4 x\, dx,$$

$$\text{or } I = \pi \int_0^{\pi/2} \sin^6 x \cos^4 x\, dx$$

$$= \pi \frac{5.3.1.3.1.}{10.8.6.4.2} \cdot \frac{\pi}{2} = \frac{3\pi^2}{512}, \qquad \text{(by Walli's formula)}$$

Example 28:

Evaluate the following integrals:

(i) $\int_0^{\pi} \sin^3 \theta (1 + 2\cos\theta)(1 + \cos\theta)^2 d\theta$

(ii) $\int_0^{\pi} \sin^5 x (1 - \cos x)^3 dx$.

Solution:

(i) Let $I = \int_0^{\pi} \sin^3 \theta (1 + 2\cos\theta)(1 + \cos\theta)^2 d\theta$

$$= \int_0^{\pi} \sin^3 \theta (1 + 2\cos\theta)\left(1 + 2\cos\theta + \cos^2\theta\right) d\theta$$

$$= \int_0^{\pi} \sin^3 \theta \left(1 + 4\cos\theta + 5\cos^2\theta + 2\cos^3\theta\right) d\theta$$

$$= \int_0^{\pi} \left(\sin^3\theta + 4\sin^3\theta \cos\theta + 5\sin^3\theta \cos^2\theta + 2\sin^3\theta \cos^3\theta\right) d\theta.$$

Now $\int_0^{\pi} \sin^m\theta \cos^n\theta\, d\theta = 2\int_0^{\pi/2} \sin^m\theta \cos^n\theta\, d\theta$ or $= 0$,

according as n is an even or an odd integer. [Refer prop. 6]

$$\therefore\ I = 2\int_0^{\pi/2} \sin^m\theta\, d\theta + 5 \times 2\int_0^{\pi/2} \sin^3\theta \cos^2\theta\, d\theta,$$

because the integrals containing odd powers of cos θ vanish

$$= 2 \cdot \frac{2}{3.1} + 10 \cdot \frac{2.1}{5.3.1} = \frac{4}{3} + \frac{4}{3} = \frac{8}{3}.$$

(ii) Let $I = \int_0^{\pi} \sin^5 x (1 - \cos x)^3 dx$

$$= \int_0^{\pi} \left(2\sin\frac{x}{2}\cos\frac{x}{2}\right)^5 \left(2\sin^2\frac{x}{2}\right)^3 dx$$

$$= 2^8 \int_0^{\pi} \sin^{11} \frac{x}{2} \cos^5 \frac{x}{2} dx.$$

Now put $\frac{x}{2} = t$,

so that dx = 2dt.

When x = 0, t = 0

and when x = π, $t = \frac{1}{2}\pi$.

$$\therefore\ I = 2\times 2^8 \int_0^{\pi/2} \sin^{11} t \cos^5 t\, dt$$

$$= 2\times 2^8 \cdot \frac{10.8.6.4.2.4.2}{16.14.12.10.8.6.4.2} = \frac{32}{21}.$$

Example 29:

Show that $\int_0^{\pi/2} \log(\tan x) dx = 0$.

Solution:

Let $I = \int_0^{\pi/2} \log(\tan x) dx = \int_0^{\pi/2} \log \tan\left(\frac{1}{2}\pi - x\right) dx,$

$$= \int_0^{\pi/2} \log \cot x\, dx = \int_0^{\pi/2} \log(\tan x)^{-1} dx$$

$$= -\int_0^{\pi/2} \log \tan x\, dx = -I.$$

∴ 2I = 0 *i.e.*, I = 0.

Example 30:

Show that $\int_0^{\pi/2} \log \sin x\, dx = -\frac{1}{2}\pi \log 2$ *or* $\frac{1}{2}\pi \log \frac{1}{2}$.

Solution:

Let $I = \int_0^{\pi/2} \log \sin x\, dx.$ ...(1)

Then $I = \int_0^{\pi/2} \log \sin\left(\frac{1}{2}\pi - x\right) dx,$ $\left[\because \int_0^a f(x) dx = \int_0^a f(a-x) dx\right]$

$$= \int_0^{\pi/2} \log \cos x\, dx. \quad ...(2)$$

Adding (1) and (2), we get

$$2I = \int_0^{\pi/2} \log \sin x\, dx + \int_0^{\pi/2} \log \cos x\, dx$$

$$= \int_0^{\pi/2} \log(\sin x \cos x) dx \quad \textbf{(Note)}$$

$$= \int_0^{\pi/2} \log\left\{\frac{\sin 2x}{2}\right\} dx = \int_0^{\pi/2} (\log \sin 2x - \log 2)\, dx$$

$$= \int_0^{\pi/2} \log \sin 2x\, dx - \int_0^{\pi/2} \log 2\, dx$$

$$= \int_0^{\pi/2} \log \sin 2x\, dx - (\log 2)\,[x]_0^{\pi/2}$$

$$= \int_0^{\pi/2} \log \sin 2x\, dx - \frac{\pi}{2} \log 2.$$

Now put $2x = t$,

so that $2dx = dt$.

Also $t = 0$ when $x = 0$

and $t = \pi$ when $x = \frac{1}{2}\pi$.

$$\therefore\ 2I = \frac{1}{2}\int_0^{\pi} \log \sin t\, dt - \frac{\pi}{2} \log 2$$

$$= \frac{1}{2}\int_0^{\pi} \log \sin x\, dx - \frac{\pi}{2} \log 2, \qquad \text{(Refer prop. 1)}$$

$$= \frac{1}{2} \cdot 2\int_0^{\pi/2} \log \sin x\, dx - \frac{\pi}{2} \log 2, \qquad \text{(Refer prop. 6)}$$

$$= I - \frac{1}{2}\pi \log 2.$$

Therefore $2I - I = -\frac{1}{2}\pi \log 2$

or $I = -\frac{1}{2}\pi \log 2 = \frac{1}{2}\pi \log(2)^{-1} = \frac{1}{2}\pi \log\frac{1}{2}$.

Example 31:

Prove that $\int_0^1 \log \sin\left(\frac{\pi}{2} y\right) dy = \log\frac{1}{2}$.

Solution:

$$u = \int_0^1 \log \sin\left(\frac{\pi}{2} y\right) dy.$$

Put $\frac{1}{2}\pi y = x$,

so that $\frac{1}{2}\pi\, dy = dx$.

When $y = 0$, $x = 0$

and when $y = 1$, $x = \frac{1}{2}\pi$.

$\therefore\ u = \int_0^{\pi/2} (\log \sin x)\cdot\frac{2}{\pi}dx = \frac{2}{\pi}\int_0^{\pi/2} \log \sin x\, dx.$

Now let $I = \int_0^{\pi/2} \log \sin x\, dx$.

We get

$I = \frac{1}{2}\pi \log\frac{1}{2}.$

$\therefore\ u = \frac{2}{\pi}I = \frac{2}{\pi}\cdot\frac{1}{2}\pi \log\frac{1}{2} = \log\frac{1}{2}.$

Example 32:

Evaluate $\int_0^{\pi} x \log \sin x\, dx$.

Solution:

Let $I = \int_0^{\pi} x \log \sin x\, dx$.

Then $I = \int_0^{\pi} (\pi - x) \log \sin(\pi - x) dx,$ (Refer prop. 4)

$= \int_0^{\pi} (\pi - x) \log \sin x\, dx$

$= \int_0^{\pi} \pi \log \sin x\, dx - \int_0^{\pi} x \log \sin x\, dx$

$= \pi \int_0^{\pi} \log \sin x\, dx - I.$

$\therefore\ 2I = \pi \int_0^{\pi} \log \sin x\, dx = 2\pi \int_0^{\pi/2} \log \sin x\, dx,$ (Refer prop. 6)

or $I = \pi \int_0^{\pi/2} \log \sin x\, dx$.

Now let $u = \int_0^{\pi/2} \log \sin x\, dx$.

We have $u = \frac{1}{2}\pi \log\frac{1}{2}.$

$\therefore\ I = \pi u = \pi.\frac{1}{2}\pi \log\frac{1}{2} = \frac{1}{2}\pi^2 \log\frac{1}{2}.$

Example 33:

Evaluate $\int_0^{\pi/2} \log \cos x\, dx$.

Solution:

Let $I = \int_0^{\pi/2} \log \cos x \, dx$. ...(1)

Then $I = \int_0^{\pi/2} \log \cos\left(\frac{1}{2}\pi - x\right) dx$ (Refer prop. 4)

$= \int_0^{\pi/2} \log \sin x \, dx$. ...(2)

Adding (1) and (2), we get

$$2I = \int_0^{\pi/2} (\log \cos x + \log \sin x) dx = \int_0^{\pi/2} \log(\sin x \cos x) dx.$$

and we get $I = \frac{1}{2}\pi \log \frac{1}{2}$.

Example 34:

Evaluate $\int_0^{\pi/2} \log \sin 2x \, dx$.

Solution:

Let $I = \int_0^{\pi/2} \log \sin 2x \, dx$.

Put 2x = t,

so that 2dx = dt.

Also t = 0 when x = 0

and t = π when $x = \frac{1}{2}\pi$.

$\therefore\ I = \frac{1}{2}\int_0^{\pi} \log \sin t \, dt = \frac{1}{2}.2\int_0^{\pi/2} \log \sin t \, dt$, [Refer prop. 6]

$= \int_0^{\pi/2} \log \sin t \, dt$.

Now we get $I = \frac{1}{2}\pi \log \frac{1}{2}$.

Example 35:

Show that $\int_0^{\pi/2} x \cot x \, dx = \frac{1}{2}\pi \log 2$.

Solution:

Let $I = \int_0^{\pi/2} x \cot x \, dx$. Integrating by parts taking cot x as the second function, we get

$$I = [x \log \sin x]_0^{\pi/2} - \int_0^{\pi/2} 1.\log \sin x \, dx$$

$$= \left[\frac{\pi}{2}\log 1 - \lim_{x\to 0} x \log \sin x\right] - \int_0^{\pi/2} \log \sin x \, dx$$

$$= 0 - \lim_{x\to 0} x \log \sin x = \lim_{x\to 0} \frac{\log \sin x}{1/x}, \qquad \left[\text{form } \frac{\infty}{\infty}\right]$$

$$= \lim_{x\to 0} \frac{\left(\frac{1}{\sin x}\right)\cos x}{-\frac{1}{x^2}} = \lim_{x\to 0} \frac{-x^2 \cos x}{\sin x}, \qquad \left[\text{form } \frac{\infty}{\infty}\right]$$

$$= \lim_{x\to 0} \frac{-2x \cos x + x^2 \sin x}{\cos x} = \frac{0}{1} = 0.$$

$$\therefore\ I = 0 - \int_0^{\pi/2} \log \sin x \, dx = -\int_0^{\pi/2} \log \sin x \, dx.$$

Now let $u = \int_0^{\pi/2} \log \sin x \, dx$.

Then we have $u = -\frac{1}{2}\pi \log 2$.

$\therefore\ I = -u = \frac{1}{2}\pi \log 2$.

Example 36:

Evaluate $\int_0^{\infty} \frac{\tan^{-1} x}{x\left(1+x^2\right)} dx$.

Solution:

$$I = \int_0^{\infty} \frac{\tan^{-1} x}{x\left(1+x^2\right)} dx.$$

Put $\tan^{-1} x = t$,

so that $\left\{\frac{1}{\left(1+x^2\right)}\right\} dx = dt$

and $x = \tan t$.
Also when $x = 0$, $t = 0$

and when $x = \infty$, $t = \frac{\pi}{2}$.

$$\therefore\ I = \int_0^{\pi/2} \frac{t\,dt}{\tan t} = \int_0^{\pi/2} t \cot t \, dt.$$

Now we get $I = \frac{1}{2}\pi \log 2$.

Example 37:

Show that $\int_0^{\pi/2}\left(\frac{\theta}{\sin\theta}\right)^2 d\theta = \pi\log 2$.

Solution:

Let $I = \int_0^{\pi/2}\left(\frac{\theta}{\sin\theta}\right)^2 d\theta = \int_0^{\pi/2}\theta^2\,\mathrm{cosec}^2\theta\, d\theta$.

Integrating by parts taking $\mathrm{cosec}^2\,\theta$ as the second function, we get

$$I = \left[\theta^2(-\cot\theta)\right]_0^{\pi/2} - \int_0^{\pi/2} 2\theta.(-\cot\theta)\,d\theta$$

$$= -\left(\frac{\pi}{2}\right)^2 \cot\frac{\pi}{2} + \lim_{\theta\to 0}\theta^2\cot\theta + 2\int_0^{\pi/2}\theta\cot\theta\,d\theta$$

$$= 0 + \lim_{\theta\to 0}\theta^2\cot\theta = \lim_{\theta\to 0}\frac{\theta^2}{\tan\theta}, \qquad \left[\text{form}\,\frac{0}{0}\right]$$

$$= \lim_{\theta\to 0}\frac{2\theta}{\sec^2\theta} = \frac{0}{1} = 0.$$

$$\therefore\ I = 2\int_0^{\pi/2}\theta\cot\theta\,d\theta.$$

Example 38:

Evaluate $\int_0^{\infty}\left(\cot^{-1}x\right)^2 dx$.

Solution:

Let $I = \int_0^{\infty}\left(\cot^{-1}x\right)^2 dx$.

Put $\cot^{-1}x = \theta$

i.e., $x = \cot\theta$,

so that $dx = -\,\mathrm{cosec}^2\,\theta\, d\theta$.

The new limits for θ are $\frac{1}{2}\pi$ to 0.

$$\therefore\ I = \int_{\pi/2}^{0}\theta^2.\left(-\mathrm{cosec}^2\theta\right)d\theta = \int_0^{\pi/2}\theta^2\,\mathrm{cosec}^2\theta\,d\theta.$$

Example 39:

Show that $\int_0^{1}\frac{\sin^{-1}x}{x}dx = \frac{1}{2}\pi\log 2$.

Solution:

Let $I = \int_0^{1}\frac{\sin^{-1}x}{x}dx$.

Put x = sin t,

so that dx = cos t dt.

Also t = 0 when x = 0

and $t = \frac{1}{2}\pi$ when x = 1.

$$\therefore\ I = \int_0^{\pi/2} \frac{t}{\sin t}\cos t\,dt = \int_0^{\pi/2} t\cot t\,dt.$$

Example 40:

Show that $\int_0^{\pi} \log(1+\cos x)\,dx = \pi\log\frac{1}{2}$.

Solution:

$$\text{Let } I = \int_0^{\pi} \log(1+\cos x)\,dx. \qquad \text{...(1)}$$

$$\text{Then } I = \int_0^{\pi} \log\{1+\cos(\pi - x)\}\,dx, \qquad \text{(Refer prop. 4)}$$

$$= \int_0^{\pi} \log(1-\cos x)\,dx. \qquad \text{...(2)}$$

Adding (1) and (2), we get

$$2I = \int_0^{\pi} [\log(1+\cos x)+\log(1-\cos x)]\,dx$$

$$= \int_0^{\pi} \log\{(1+\cos x)\,(1-\cos x)\}$$

$$= \int_0^{\pi} \log\left(1-\cos^2 x\right)dx = \int_0^{\pi} \log\sin^2 x\,dx = 2\int_0^{\pi} \log\sin x\,dx.$$

$$\therefore\ I = \int_0^{\pi} \log\sin x\,dx = 2\int_0^{\pi/2} \log\sin x\,dx. \qquad \text{(Refer prop. 6)}$$

Now let $u = \int_0^{\pi/2} \log\sin x\,dx$.

We have $u = \frac{1}{2}\pi\log\frac{1}{2}$.

$$\therefore\ I = 2u = 2.\frac{1}{2}\pi\log\frac{1}{2} = \pi\log\frac{1}{2}.$$

Example 41:

Show that $\int_0^{\infty} \log\left(x+\frac{1}{x}\right)\frac{dx}{1+x^2} = \pi\log 2$.

Solution:

Put x = tan θ, so that dx = $\sec^2\theta\,d\theta$.

Also when x = 0, θ = 0

and when $x \to \infty$, $\theta \to \frac{\pi}{2}$.

$$\therefore\ I = \int_0^{\infty} \log\left(x + \frac{1}{x}\right) \frac{dx}{1+x^2}$$

$$= \int_0^{\pi/2} \log\left(\tan\theta + \frac{1}{\tan\theta}\right) \frac{\sec^2\theta}{\sec^2\theta} d\theta$$

$$= \int_0^{\pi/2} \log\left(\frac{1+\tan^2\theta}{\tan\theta}\right) d\theta = \int_0^{\pi/2} \log\frac{\sec^2\theta}{\tan\theta} d\theta$$

$$= \int_0^{\pi/2} \log\left(\frac{1}{\sin\theta\cos\theta}\right) d\theta = \int_0^{\pi/2} \log(\sin\theta\cos\theta)^{-1} d\theta$$

$$= -\int_0^{\pi/2} \log\sin\theta\, d\theta - \int_0^{\pi/2} \log\cos\theta\, d\theta.$$

Now let $u = \int_0^{\pi/2} \log\sin\theta\, d\theta$.

We have $u = \int_0^{\pi/2} \log\cos\theta\, d\theta = -\frac{\pi}{2}\log 2$.

$$\therefore\ I = -u - u = -2u = -2\left(-\frac{1}{2}\pi\log 2\right) = \pi\log 2.$$

Example 42:

Show that $\int_0^{\infty} \frac{\log\left(1+x^2\right)dx}{\left(1+x^2\right)} = \pi\log 2$.

Solution:

Put x = tan θ,

so that dx = $\sec^2\theta\, d\theta$.

The new limits for θ are 0 to $\frac{1}{2}\pi$.

$$\therefore\ I = \int_0^{\infty} \frac{\log\left(1+x^2\right)}{\left(1+x^2\right)} dx = \int_0^{\pi/2} \frac{\log\left(1+\tan^2\theta\right)\sec^2\theta\, d\theta}{\sec^2\theta}$$

$$= \int_0^{\pi/2} \log\sec^2\theta\, d\theta, \qquad (\because\ 1 + \tan^2\theta = \sec^2\theta)$$

$$= 2\int_0^{\pi/2} \log\sec\theta\, d\theta = 2\int_0^{\pi/2} \log(\cos\theta)^{-1} d\theta$$

$$= -2\int_0^{\pi/2} \log\cos\theta\, d\theta.$$

Now let $u = \int_0^{\pi/2} \log\cos\theta\, d\theta$.

We have $u = -\frac{1}{2}\pi \log 2$.

$\therefore\ I = -2u = -2.\left(-\frac{1}{2}\pi \log 2\right) = \pi \log 2$.

Example 43:

Show that $\int_0^{\pi/4} \log(1+\tan\theta)d\theta = \frac{\pi}{8}\log 2$.

Solution:

Let $I = \int_0^{\pi/4} \log(1+\tan\theta)d\theta$.

Then $I = \int_0^{\pi/4} \log\left\{1+\left(\frac{1}{4}\pi - \theta\right)\right\}d\theta$,

$\left[\because \int_0^a f(x)dx = \int_0^a f(a-x)dx\right]$

$= \int_0^{\pi/4} \log\left[1+\frac{(1-\tan\theta)}{(1+\tan\theta)}\right]d\theta = \int_0^{\pi/4} \log\left\{\frac{2}{1+\tan\theta}\right\}d\theta$

$= \int_0^{\pi/4} \log 2.d\theta - \int_0^{\pi/4} \log(1+\tan\theta)d\theta$

$= \log 2.[\theta]_0^{\pi/4} - I$.

$\therefore\ 2I = \frac{1}{4}\pi \log 2$

or $\quad I = \frac{1}{8}\pi \log 2$.

Example 44:

Show that $\int_0^1 \frac{\log(1+x)}{1+x^2}dx = \frac{1}{8}\pi\log 2$.

Solution:

Put x = tan θ,

so that $dx = \sec^2\theta\, d\theta$.

And the new limits are θ = 0 to $\theta = \frac{\pi}{4}$.

$\therefore\ I = \int_0^1 \frac{\log(1+x)}{1+x^2}dx = \int_0^{\pi/4} \frac{\log(1+\tan\theta)}{\sec^2\theta}\sec^2\theta\, d\theta$

$$= \int_0^{\pi/4} \log(1+\tan\theta)d\theta = \frac{1}{8}\pi \log 2.$$

Example 45:

Show that $\int_0^{\pi/2} \frac{\sin x\, dx}{\sin x + \cos x} = \frac{\pi}{4}$.

Solution:

Let $I = \int_0^{\pi/2} \frac{\sin x\, dx}{\sin x + \cos x}$. ...(1)

Then $I = \int_0^{\pi/2} \frac{\sin\left(\frac{1}{2}\pi - x\right)}{\sin\left(\frac{1}{2}\pi - x\right) + \cos\left(\frac{1}{2}\pi - x\right)}$ (Refer prop. 4)

$$= \int_0^{\pi/2} \frac{\cos x\, dx}{\cos x + \sin x}. \qquad ...(2)$$

Adding (1) and (2), we get

$$2I = \int_0^{\pi/2} \frac{\sin x\, dx}{\sin x + \cos x} + \int_0^{\pi/2} \frac{\cos x\, dx}{\sin x + \cos x}$$

$$= \int_0^{\pi/2} \left[\frac{\sin x}{\sin x + \cos x} + \frac{\cos x}{\sin x + \cos x}\right] dx$$

$$= \int_0^{\pi/2} 1.dx = [x]_0^{\pi/2} = \frac{\pi}{2}.$$

$$\therefore\ I = \frac{1}{4}\pi.$$

Similarly we can prove that

$$\int_0^{\pi/2} \frac{\cos x\, dx}{\sin x + \cos x} = \frac{\pi}{4}.$$

Example 46:

Evaluate $\int_0^{\pi/2} \frac{dx}{1 + \tan x}$.

Solution:

We have $I = \int_0^{\pi/2} \frac{dx}{1 + \tan x} = \int_0^{\pi/2} \frac{dx}{1 + \left(\frac{\sin x}{\cos x}\right)}$

$$= \int_0^{\pi/2} \frac{\cos x\, dx}{\cos x + \sin x}$$

$$= \frac{\pi}{4}.$$

Example 47:

Evaluate $\int_0^{\pi/2} \frac{dx}{1+\cot x}$.

Solution:

We have $\int_0^{\pi/2} \frac{dx}{1+\cot x}$

$$= \int_0^{\pi/2} \frac{dx}{1+\left(\frac{\cos x}{\sin x}\right)} = \int_0^{\pi/2} \frac{\sin x\, dx}{\sin x + \cos x}$$

$$= \frac{\pi}{4}.$$

Example 48:

Show that $\int_0^{\infty} \frac{x\, dx}{(1+x)\left(1+x^2\right)} = \frac{\pi}{4}$.

Solution:

Put x = tan θ,

so that dx = $\sec^2 \theta\, d\theta$.

Also when x = 0, θ = 0

and when $x \to \infty$, $\theta \to \frac{\pi}{2}$.

∴ the given integral $I = \int_0^{\pi/2} \frac{\tan\theta \sec^2\theta\, d\theta}{(1+\tan\theta)\left(1+\tan^2\theta\right)}$

$$= \int_0^{\pi/2} \frac{\tan\theta\, d\theta}{1+\tan\theta} = \int_0^{\pi/2} \frac{\frac{\sin\theta}{\cos\theta}}{1+\left(\frac{\sin\theta}{\cos\theta}\right)} d\theta$$

$$= \int_0^{\pi/2} \frac{\sin\theta\, d\theta}{\cos\theta + \sin\theta} = \frac{\pi}{4}.$$

Example 49:

Show that $\int_0^{a} \frac{dx}{x+\sqrt{\left(a^2-x^2\right)}} = \frac{\pi}{4}$.

Solution:

Put x = a sin θ,

so that dx = a cos θ dθ.

Also the new limits are θ = 0 to $\theta = \frac{\pi}{2}$.

$\therefore$ the given integral $I = \int_0^{\pi/2} \frac{a\cos\theta\, d\theta}{a\sin\theta + a\cos\theta}$

$$= \int_0^{\pi/2} \frac{\cos\theta\, d\theta}{\sin\theta + \cos\theta} = \frac{\pi}{4}.$$

Example 50:

Show that $\int_0^{\pi/2} \frac{\sqrt{(\sin x)}}{\sqrt{(\sin x)} + \sqrt{(\cos x)}} dx = \frac{\pi}{4}.$

Solution:

Let $I = \int_0^{\pi/2} \frac{\sqrt{(\sin x)}}{\sqrt{(\sin x)} + \sqrt{(\cos x)}} dx.$ ...(1)

Then $I = \int_0^{\pi/2} \frac{\sqrt{\left[\sin\left(\frac{1}{2}\pi - x\right)\right]} dx}{\sqrt{\left[\sin\left(\frac{1}{2}\pi - x\right)\right]} + \sqrt{\left[\cos\left(\frac{1}{2}\pi - x\right)\right]}}$, (Refer prop. 4).

$$= \int_0^{\pi/2} \frac{\sqrt{(\cos x)}\, dx}{\sqrt{(\cos x)} + \sqrt{(\sin x)}}. \quad ...(2)$$

Adding (1) and (2), we get

$$2I = \int_0^{\pi/2} \left[\frac{\sqrt{(\sin x)}}{\sqrt{(\sin x)} + \sqrt{(\cos x)}} + \frac{\sqrt{(\cos x)}}{\sqrt{(\sin x)} + \sqrt{(\cos x)}}\right] dx$$

$$= \int_0^{\pi/2} \frac{\sqrt{(\sin x)} + \sqrt{(\cos x)}}{\sqrt{(\sin x)} + \sqrt{(\cos x)}} dx = \int_0^{\pi/2} 1.dx = \frac{\pi}{2}.$$

$$\therefore I = \frac{1}{4}\pi.$$

Example 51:

Show that $\int_0^{\pi/2} \frac{dx}{1+\sqrt{(\tan x)}} = \frac{\pi}{4}.$

Solution:

We have

$$I = \int_0^{\pi/2} \frac{dx}{1+\sqrt{(\tan x)}} = \int_0^{\pi/2} \frac{dx}{1+\sqrt{\left(\frac{\sin x}{\cos x}\right)}}$$

$$= \int_0^{\pi/2} \frac{\sqrt{(\cos x)}\,dx}{\sqrt{(\cos x)} + \sqrt{(\sin x)}}.$$

Example 52:

Show that $\int_0^{\pi/2} \frac{\sqrt{(\tan x)}\,dx}{1+\sqrt{(\tan x)}} = \frac{\pi}{4}.$

Solution:

We have

$$I = \int_0^{\pi/2} \frac{\sqrt{(\tan x)}\,dx}{1+\sqrt{(\tan x)}} = \int_0^{\pi/2} \frac{\sqrt{(\sin x)}\,dx}{\sqrt{(\cos x)} + \sqrt{(\sin x)}} \qquad \left(\because \tan x = \frac{\sin x}{\cos x}\right)$$

$$= \frac{\pi}{4}.$$

Example 53:

Prove that $\int_0^{\pi/2} \frac{\sqrt{(\tan x)}}{\sqrt{(\tan x)} + \sqrt{(\cot x)}}\,dx = \frac{\pi}{4}.$

Solution:

Here $I = \int_0^{\pi/2} \frac{\sin x}{\sin x + \cos x}\,dx$, changing to sin x and cos x.

Example 54:

Show that $\int_0^{\pi/2} \frac{\sin^2 x\,dx}{(\sin x + \cos x)} = \frac{1}{\sqrt{2}} \log\left(\sqrt{2}+1\right).$

Solution:

Let $I = \int_0^{\pi/2} \frac{\sin^2 x\,dx}{(\sin x + \cos x)}.$ ...(1)

Then $I = \int_0^{\pi/2} \frac{\left[\sin\left(\frac{1}{2}\pi - x\right)\right]^2 dx}{\sin\left(\frac{1}{2}\pi - x\right) + \cos\left(\frac{1}{2}\pi - x\right)},$ (Refer prop. 4)

or $I = \int_0^{\pi/2} \frac{\cos^2 x\,dx}{\cos x + \sin x}.$ ...(2)

Adding (1) and (2), we get

$$2I = \int_0^{\pi/2} \frac{\sin^2 x\,dx}{\sin x + \cos x} + \int_0^{\pi/2} \frac{\cos^2 x\,dx}{\cos x + \sin x}$$

$$= \int_0^{\pi/2} \frac{(\sin^2 x + \cos^2 x)}{(\sin x + \cos x)} dx = \int_0^{\pi/2} \frac{dx}{(\sin x + \cos x)}$$

$$= \int_0^{\pi/2} \frac{\left(\frac{1}{\sqrt{2}}\right) dx}{\left(\frac{1}{\sqrt{2}}\right)\sin x + \left(\frac{1}{\sqrt{2}}\right)\cos x} \qquad \textbf{(Note)}$$

$$= \frac{1}{\sqrt{2}} \int_0^{\pi/2} \frac{dx}{\cos\left(x - \frac{1}{4}\pi\right)}, \qquad \left[\because \cos\frac{\pi}{4} = \sin\frac{\pi}{4} = \frac{1}{\sqrt{2}}\right]$$

$$= \frac{1}{\sqrt{2}} \int_0^{\pi/2} \sec\left(x - \frac{1}{4}\pi\right) dx$$

$$= \frac{1}{\sqrt{2}} \log\left[\sec\left(x - \frac{\pi}{4}\right) + \tan\left(x - \frac{\pi}{4}\right)\right]_0^{\pi/2}$$

$$= \frac{1}{\sqrt{2}}\left[\log\left(\sec\frac{1}{4}\pi + \tan\frac{1}{4}\pi\right) - \log\left\{\sec\left(-\frac{1}{4}\pi\right) + \tan\left(-\frac{1}{4}\pi\right)\right\}\right]$$

$$= \frac{1}{\sqrt{2}} \log\left[\frac{\sec\frac{1}{4}\pi + \tan\frac{1}{4}\pi}{\sec\left(-\frac{1}{4}\pi\right) + \tan\left(-\frac{1}{4}\pi\right)}\right] = \frac{1}{\sqrt{2}} \log\left[\frac{\sqrt{2}+1}{\sqrt{2}-1}\right]$$

$$= \frac{1}{\sqrt{2}} \log\left[\frac{(\sqrt{2}+1)(\sqrt{2}+1)}{(\sqrt{2}-1)(\sqrt{2}+1)}\right] = \frac{1}{\sqrt{2}} \log(\sqrt{2}+1)^2$$

$$= \left(\frac{1}{\sqrt{2}}\right).2\log(\sqrt{2}+1).$$

$$\therefore\ I = \frac{1}{\sqrt{2}} \log(\sqrt{2}+1).$$

Example 55:

Evaluate $\int_0^{\pi/2} \frac{\cos^2 x}{(\sin x + \cos x)} dx$.

Solution:

Do your self.

Example 56:

Evaluate $\int_0^{a} \frac{a\,dx}{\left\{x + \sqrt{(a^2 - x^2)}\right\}^2}$.

Solution:

Put $x = a \sin \theta$,

so that $dx = a \cos \theta \, d\theta$.

When $x = 0$, $\theta = 0$

and when $x = a$, $\theta = \dfrac{\pi}{2}$.

$\therefore$ the given integral $I = \displaystyle\int_0^{\pi/2} \frac{a.a \cos\theta \, d\theta}{a^2(\sin\theta + \cos\theta)^2}$

$$= \int_0^{\pi/2} \frac{\cos\theta \, d\theta}{(\sin\theta + \cos\theta)^2}. \qquad ...(1)$$

$$\text{Also } I = \int_0^{\pi/2} \frac{\cos\left(\frac{1}{2}\pi - \theta\right) d\theta}{\left[\sin\left(\frac{1}{2}\pi - \theta\right) + \cos\left(\frac{1}{2}\pi - \theta\right)\right]^2}, \qquad \text{(Refer prop. 4)}$$

$$= \int_0^{\pi/2} \frac{\sin\theta \, d\theta}{(\cos\theta + \sin\theta)^2}. \qquad ...(2)$$

Adding (1) and (2), we get

$$2I = \int_0^{\pi/2} \frac{\cos\theta + \sin\theta}{(\sin\theta + \cos\theta)^2} d\theta = \int_0^{\pi/2} \frac{d\theta}{\sin\theta + \cos\theta}$$

$$= \left(\frac{1}{\sqrt{2}}\right).2 \log\left(\sqrt{2} + 1\right),$$

$$\therefore \; I = \left(\frac{1}{\sqrt{2}}\right) \log\left(\sqrt{2} + 1\right).$$

Example 57:

Evaluate $\displaystyle\int_0^{\pi/2} \frac{x \, dx}{\sin x + \cos x}$.

Solution:

$$\text{Let } I = \int_0^{\pi/2} \frac{\left(\frac{1}{2}\pi - x\right) dx}{\sin\left(\frac{1}{2}\pi - x\right) + \cos\left(\frac{1}{2}\pi - x\right)}, \qquad ...(1)$$

$$\text{Then } I = \int_0^{\pi/2} \frac{\left(\frac{1}{2}\pi - x\right) dx}{\sin\left(\frac{1}{2}\pi - x\right) + \cos\left(\frac{1}{2}\pi - x\right)}. \qquad \text{(Refer prop. 4)}$$

$$= \int_0^{\pi/2} \frac{\left(\frac{1}{2}\pi - x\right)dx}{\sin x + \cos x}. \qquad ...(2)$$

Adding (1) and (2), we get

$$2I = \int_0^{\pi/2} \frac{\frac{1}{2}\pi\, dx}{\sin x + \cos x} = \frac{\pi}{2}\int_0^{\pi/2} \frac{dx}{\sin x + \cos x}.$$

$$\therefore\ I = \frac{\pi}{4}\int_0^{\pi/2} \frac{dx}{\sin x + \cos x}.$$

Now we have

$$\int_0^{\pi/2} \frac{dx}{\sin x + \cos x} = \frac{1}{\sqrt{2}}\cdot 2\log\left(\sqrt{2}+1\right).$$

$$\therefore\ I = \frac{\pi}{4}\cdot\frac{1}{\sqrt{2}}\cdot 2\log\left(\sqrt{2}+1\right) = \frac{\pi}{2\sqrt{2}}\log\left(\sqrt{2}+1\right).$$

Example 58:

Show that

$$\int_0^{\pi/2} \phi(\sin 2x)\ \sin x\, dx = \int_0^{\pi/2} \phi(\sin 2x)\cos x\, dx$$

$$= \sqrt{2}\int_0^{\pi/4} \phi(\cos 2x)\cos x\, dx.$$

Solution:

We have $\int_0^{\pi/2} \phi(\sin 2x).\sin x\, dx$

$$= \int_0^{\pi/2} \phi\left[\sin 2\left(\frac{\pi}{2} - x\right)\right]\cdot\sin\left(\frac{\pi}{2} - x\right)dx, \qquad \text{(Refer prop. 4)}$$

$$d= \int_0^{\pi/2} \phi[\sin(\pi - 2x)]\cos x\, dx$$

$$= \int_0^{\pi/2} \phi(\sin 2x)\cos x\, dx.$$

Now to prove the second part, let

$$I = \int_0^{\pi/2} \phi(\sin 2x)\sin x\, dx.$$

Put $x = \frac{1}{4}\pi + t$,

so that dx = dt.

When $x = 0,\ t = -\frac{\pi}{4}$

and when $x = \frac{\pi}{2},\ t = \frac{\pi}{4}$.

$$\therefore\ I = \int_{-\pi/4}^{\pi/4} \phi\left[\sin 2\left(\frac{1}{4}\pi + t\right)\right] \sin\left(\frac{1}{4}\pi + t\right) dt$$

$$= \int_{-\pi/4}^{\pi/4} \phi\left[\sin\left(\frac{1}{2}\pi + 2t\right)\right] \sin\left(\frac{1}{4}\pi + t\right) dt$$

$$= \int_{-\pi/4}^{\pi/4} \phi(\cos 2t)\left(\sin\frac{1}{4}\pi\ \cos t + \cos\frac{1}{4}\pi\ \sin t\right) dt$$

$$= \frac{1}{\sqrt{2}}\int_{-\pi/4}^{\pi/4} \phi(\cos 2t)\cos t\, dt + \frac{1}{\sqrt{2}}\int_{-\pi/4}^{\pi/4} \phi(\cos 2t)\sin t\, dt.$$

Now $\int_{-\pi/4}^{\pi/4} \phi(\cos 2t)\sin t\, dt = 0$, because ϕ (cos 2t) sin t is an odd function of t

and $\int_{-\pi/4}^{\pi/4} \phi(\cos 2t)\cos t\, dt = 2\int_{0}^{\pi/4} \phi(\cos 2t)\cos t\, dt$

because ϕ(cos 2t) cos t is an even function of t.

$$\therefore\ I = \frac{2}{\sqrt{2}}\int_{0}^{\pi/4} \phi(\cos 2t)\cos t\, dt$$

$$= \sqrt{2}\int_{0}^{\pi/4} \phi(\cos 2x)\cos x\, dx,$$

because a definite integral does not change by changing the variable.

Example 59:

Show that $\int_0^{\pi} \sin^m x \cos^{2m+1} x\, dx = 0$.

Solution:

Let $I = \int_0^{\pi} \sin^m x \cos^{2m+1} x\, dx$.

Then $I = \int_0^{\pi} \sin^m(\pi - x).\cos^{2m+1}(\pi - x)\, dx$, (Refer prop. 4)

$$= \int_0^{\pi} \sin^m x(-\cos x)^{2m+1} dx$$

$$= -\int_0^{\pi} \sin^m x \cos^{2m+1} x\, dx = -I.$$

$\therefore\ 2I = 0$ or $I = 0$.

Example 60:

Show that $\int_0^{\pi} \frac{x^2 \sin 2x \sin\left(\frac{1}{2}\pi \cos x\right)}{2x-\pi} dx = \frac{8}{\pi}$.

Solution:

Let $I = \int_0^{\pi} \frac{x^2 \sin 2x.\sin\left(\frac{1}{2}\pi \cos x\right)}{2x-\pi} dx$.

Put $x = \frac{1}{2}\pi - t$,

so that $dx = -dt$.

Also $t = \frac{1}{2}\pi$ when $x = 0$ a

nd $t = -\frac{1}{2}\pi$ when $x = \pi$.

$$\therefore I = \int_{\pi/2}^{-\pi/2} \frac{\left(\frac{1}{2}\pi - t\right)^2 \sin 2\left(\frac{1}{2}\pi - t\right).\sin\left\{\frac{1}{2}\pi \cos\left(\frac{1}{2}\pi - t\right)\right\}}{2\left(\frac{1}{2}\pi - t\right) - \pi}(-dt)$$

$$= \int_{-\pi/2}^{\pi/2} \frac{\left(\frac{1}{2}\pi - t\right)^2 \sin 2t.\sin\left(\frac{1}{2}\pi \sin t\right)}{-2t} dt$$

$$= -\frac{1}{2}\int_{-\pi/2}^{\pi/2} \frac{\left(\frac{1}{4}\pi^2 - \pi t + t^2\right)\sin 2t.\sin\left(\frac{1}{2}\pi \sin t\right)}{t} dt.$$

Now $\frac{\frac{1}{4}\pi^2 \sin 2t.\sin\left(\frac{1}{2}\pi \sin t\right)}{t}$ and $t \sin 2t. \sin\left(\frac{1}{2}\pi \sin t\right)$ are both odd functions of t while $\sin 2t.\sin\left(\frac{1}{2}\pi \sin t\right)$ is an even function of t.

$$\therefore\ I = -\frac{1}{2}.2\int_0^{\pi/2} (-\pi)\sin 2t.\sin\left(\frac{1}{2}\pi \sin t\right)dt, \qquad \text{(Refer prop. 5)}$$

$$= \pi \int_0^{\pi/2} 2\sin t \cos t.\sin\left(\frac{1}{2}\pi \sin t\right)dt.$$

Now put $\frac{1}{2}\pi \sin t = z$,

so that $\frac{1}{2}\pi \cos t\,dt = dz$.

Also $z = 0$ when $t = 0$

and $z = \frac{1}{2}\pi$ when $t = \frac{1}{2}\pi$.

$$\therefore\ I = \pi\int_0^{\pi/2} \frac{2.2z}{\pi}\cdot \sin z \cdot \frac{2}{\pi} dz = \frac{8}{\pi}\int_0^{\pi/2} z \sin z\, dz$$

$$= \frac{8}{\pi}\left[\{z(-\cos z)\}_0^{\pi/2} - \int_0^{\pi/2} 1.(-\cos z)dz\right]$$

$$= \frac{8}{\pi}\left[0 + \int_0^{\pi/2} \cos z\,dz\right] = \frac{8}{\pi}[\sin z]_0^{\pi/2} = \frac{8}{\pi}(1-0) = \frac{8}{\pi}.$$

Example 61:

Evaluate $\int_a^b x^2\,dx$ *directly from the definition of the integral as the lmit of a sum.*

Solution:

From the definition of a definite integral as the limit of a sum, we know that

$$\int_a^b f(x)dx = \lim_{h\to\infty} h\,[f(a) + f(a+2h + \ldots + f\{a + (n-1)h\}],$$

where h ® 0 as n ® ¥ and nh ® b – a.

Here $f(x) = x^2$; \ $f(a)$, $f(a + h)$, $f(a + 2h)$, etc. will be a^2, $(a + h)^2$, $(a + 2h)^2$, ..., respectively.

$$\therefore\ \int_a^b x^2 dx = \lim_{n\to\infty} h[a^2 + (a+h)^2 + (a+2h)^2 + \ldots$$

$$+ \{a + (n-1)h\}^2],$$

where $h \to 0$ as $n \to \infty$ and $nh \to b - a$

$$= \lim_{n\to\infty} h[na^2 + 2ah\{1 + 2 + 3 + \ldots + (n-1)\}$$

$$+ h^2\{1^2 + 2^2 + 3^2 + \ldots + (n-1)^2\}].$$

But we know that

$$\Sigma n = \frac{n(n-1)}{2}$$

and $\Sigma n^2 = \frac{n(n+1)(2n+1)}{6}$.

Taking n = (n – 1) in the above results, we get

$$\int_a^b x^2 dx = \lim_{n\to\infty} h\left[na^2 + 2ah\cdot\frac{(n-1)n}{2} + \frac{h^2}{6}(n-1)n(2n-1)\right]$$

$$= \lim_{n\to\infty}\left[(nh)a^2 + a(nh)(n-1)h + \frac{1}{6}(nh)(n-1)h(2n-1)h\right]$$

$$= \lim_{n\to\infty}\left[(nh)a^2 + a(nh)^2\left(1-\frac{1}{n}\right) + \frac{1}{6}\cdot 2(nh)^3\left(1-\frac{1}{n}\right)\left(1-\frac{1}{2n}\right)\right]$$

Now as $n \to \infty$, $h \to 0$ and $nh \to b - a$.

$$\therefore \int_a^b x^2 dx = (b-a)a^2 + a(b-a)^2 + \frac{1}{3}(b-a)^3$$

$$= \frac{1}{3}(b-a)\left\{3a^2 + 3(b-a)a + b^2 - 2ab + a^2\right\}$$

$$= \frac{1}{3}(b-a)\left(a^2 + ab + b\right)^2 = \frac{1}{3}\left(b^3 - a^3\right).$$

Example 62:

Find by summation the value of $\int_a^b x\,dx$.

Solution:

Here f(x) = x;

∴ f(a) = a, f(a + h) = a + h,

f(a + 2h) = a + 2h, etc.

Now $\int_a^b f(x)dx = \lim_{h\to 0} h[f(a) + f(a+h) + \ldots + f\{a+(n-1)h\}]$,

where $n \to \infty$ and $nh \to b - a$ as $h \to 0$.

$$\therefore \int_a^b x\,dx = \lim_{h\to 0} h[a + (a+h) + (a+2h) + \ldots + \{a+(n-1)h\}]$$

$= \lim_{h\to 0} h\left[\frac{n}{2}\{2a + (n-1)h\}\right]$, summing the A.P.

$$= \lim_{h\to 0} \frac{nh}{2}[2a + nh - h]$$

$= \lim_{h\to 0} \frac{b-a}{2}[2a + (b-a) - h]$, $[\because nh = b - a]$

$$= \frac{1}{2}(b-a)\left[2a+(b-a)\right] = \frac{1}{2}(b-a)(b+a) = \frac{1}{2}\left(b^2-a^2\right).$$

Example 63:

Evaluate the summation $\int_1^2 x\,dx$.

Solution:

Do your self.

Here b = 2, a = 1.

Thus, we get

$$\int_1^2 x\,dx = \frac{1}{2}(4-1) = \frac{3}{2}.$$

Example 64:

Evaluate by summation $\int_0^2 x^3 dx$.

Solution:

Here $f(x) = x^3$ and a = 0, b = 2; $\therefore$ nh = 2 – 0 = 2.

$$\therefore \int_0^2 x^3 dx = \lim_{h\to 0} h\left[0^3 + h^3 + 2^3h^3 + 3^3h^3 + \ldots + (n-1)^3 h^3\right]$$

$$= \lim_{h\to 0} h^4\left[1^3 + 2^3 + 3^3 + \ldots + (n-1)^3\right], \text{ where } nh = 2$$

$$= \lim_{h\to 0} h^4\left[\frac{(n-1)^2\{(n-1)+1\}^2}{4}\right],$$

summaing up the series using the formula $\Sigma n^3 = \left[\frac{n(n+1)}{2}\right]^2$

$$= \lim_{h\to 0}\frac{1}{4}h^4(n-1)^2 n^2, \text{ where } nh = 2$$

$$= \lim_{h\to 0}\frac{1}{4}(nh-h)^2(nh)^2 = \frac{1}{4}(2-0)^2.2^2 = 4.$$

Example 65:

Show that $\int_a^b x^m dx = \dfrac{b^{m+1} - a^{m+1}}{(m+1)}$.

Solution:

Here $f(x) = x^m$; $\therefore$ $f(a) = a^m$,

$f(a + h) = (a + h)^m$, etc.

$$\therefore \int_a^b x^m dx = \lim_{h \to 0} h\left[a^m + (a+h)^m + \ldots + \{a+(n-1)h\}^m\right],$$

where b – a = nh.

Now $\lim_{h \to 0} \frac{(t+h)^{m+1} - t^{m+1}}{h} = \frac{d}{dt} t^{m+1} = (m+1)t^m$.

$\therefore \lim_{h \to 0} \frac{(t+h)^{m+1} - t^{m+1}}{h.t^m} = (m+1)$, *i.e.*, a constant.

Putting t = a, (a + h), (a + 2h), etc. in (1), we get

$$\lim_{h \to 0} \frac{(a+h)^{m+1} - a^{m+1}}{h.a^m} = \lim_{h \to 0} \frac{(a+2h)^{m+1} - (a+h)^{m+1}}{h(a+h)^m} = \ldots$$

$$\ldots = \lim_{h \to 0} \frac{(a+nh)^{m+1} - \{a+(n-1)h\}^{m+1}}{h\{a+(n-1)h\}^m} = (m+1)$$

i.e., a constant. ...(2)

Also we know that if $\frac{a}{b} = \frac{c}{d} = \frac{e}{f} = \ldots$,

then each of these ratios is equal to $\frac{a+c+e+\ldots}{b+d+f+\ldots}$ ...(3)

Now we apply the property (3) to various limits given in (2). Thus, forming a new numerator and denominator by adding the numerators and denominators of the various ratios in (2), we get

$$\lim_{h \to 0} \frac{(a+nh)^{m+1} - a^{m+1}}{h\left[a^m + (a+h)^m + \ldots + \{a+(n-1)h\}^m\right]} = (m+1)$$

or $$\lim_{h \to 0} \frac{\left[a+(b-a)^{m+1} - a^{m+1}\right]}{h\left[a^m + (a+h)^m + \ldots + \{a+(n-1)h\}^m\right]} = (m+1),$$

[∵ nh = b – a]

or $\lim_{h \to 0} h\left[a^m + (a+h)^m + \ldots + \{a+(n-1)h\}^m\right]$

$$= \frac{b^{m+1} - a^{m+1}}{m+1}. \therefore \int_a^b x^m dx = \frac{b^{m+1} - a^{m+1}}{(m+1)}.$$

Example 66:

From the definition of a definite integral as the limit of a sum, evaluate $\int_a^b e^x dx$.

Solution:

Here $f(x) = e^x$; $\therefore$ $f(a) = e^a$, $f(a + h) = e^{a+h}$, etc.

$$\therefore \int_a^b e^x dx = \lim_{h\to 0} h\{e^a + e^{a+h} + e^{a+2h} + \ldots + e^{a+(n-1)h}\},$$

where $nh = b - a$

and $n \to \infty$ as $h \to 0$

$$= \lim_{h\to 0} he^a \{1 + e^h + e^{2h} + \ldots + e^{(n-1)h}\}$$

$$= \lim_{h\to 0} he^a \left\{\frac{(e^h)^n - 1}{e^h - 1}\right\}, \text{ summing the G.P.}$$

$$= \lim_{h\to 0} he^a \left[\frac{e^{nh} - 1}{e^h - 1}\right] = \lim_{h\to 0} he^a \left[\frac{e^{b-a} - 1}{e^h - 1}\right], \qquad [\because nh = (b - a)]$$

$$= e^a (e^{b-a} - 1), \qquad \left\{\because \lim_{h\to 0}\frac{h}{e^h - 1} = \lim_{h\to 0}\frac{1}{e - 1}\right\}$$

$$= e^b - e^a.$$

Example 67:

Evaluate by summation $\int_a^b \sin x\, dx$.

Solution:

Here $f(x) = \sin x$; $\therefore$ $f(a) = \sin a$,

$f(a + h) = \sin (a + h)$, etc.

$$\therefore \int_a^b \sin x\, dx = \lim_{h\to 0} h[\sin a + \sin(a + h) + \ldots + \ldots \sin\{a + (n - 1)h\}],$$

where $nh = b - a$ and $n \to \infty$ as $h \to 0$

$$= \lim_{h\to 0} h\left[\frac{\sin\left(\frac{1}{2}nh\right)}{\sin\frac{1}{2}h}.\sin\left\{a + \frac{1}{2}(n - 1)h\right\}\right], \qquad \text{(from Trigonometry)}$$

$$= \lim_{h\to 0} 2.\frac{\frac{1}{2}h}{\sin\frac{1}{2}h}.\sin\left(\frac{b-a}{2}\right).\sin\left(a + \frac{b-a-h}{2}\right) \qquad [\because nh = b - a]$$

$$= 2.1.\sin\left(\frac{b-a}{2}\right)\sin\left(a + \frac{b-a}{2}\right), \qquad \left\{\because \lim_{\theta\to 0}\frac{\theta}{\sin\theta} = 1\right\}$$

$$= 2\sin\frac{b-a}{2}\sin\frac{a+b}{2} = \cos a - \cos b.$$

Example 68:

Using the definition of integral as the limit of a sum, show that $\int_a^b \cos x \, dx - \sin b - \sin a$.

Solution:

Do your self.

Example 69(c):

Evaluate by summation $\int_0^{\pi/2} \sin x \, dx$.

Solution:

Here f(x) = sin x;

a = 0 and $b = \frac{\pi}{2}$,

$nh = b - a = \frac{1}{2}\pi - 0 = \frac{1}{2}\pi$.

We get $\int_0^{\pi/2} \sin x \, dx = \cos 0 - \cos \frac{1}{2}\pi = 1 - 0 = 1$.

Example 70:

Evaluate by summation $\int_0^{\pi/2} \cos x \, dx$.

Solution:

Here f(x) = cos x; a = 0

and $b = \frac{\pi}{2}$,

$nh = b - a = \frac{1}{2}\pi - 0 = \frac{1}{2}\pi$.

We get $\int_0^{\pi/2} \cos x \, dx = \sin \frac{1}{2}\pi - \sin 0 = 1 - 0 = 1$.

Example 71:

Evaluate the summation $\int_a^b \frac{1}{x^2} dx$.

Solution:

Do your self.

Put m = – 2.

Example 72:

Show that the limit of the sum

$$\frac{1}{n}+\frac{1}{n+1}+\frac{1}{n+2}+...+\frac{1}{3n},$$

when n is indefinitely increased is log 3.

Solution:

Here the general term of the series is $\frac{1}{n+r}$ and r varies from 0 to 2n.

Now we have to find $\lim_{n\to\infty}\sum_{r=0}^{2n}\frac{1}{n+r}$.

We have $\lim_{n\to\infty}\sum_{r=0}^{2n}\frac{1}{n+r}=\lim_{n\to\infty}\sum_{r=0}^{2n}\frac{1}{b\left\{1+\left(\frac{r}{n}\right)\right\}}$, expressing the general term in the form $\left(\frac{1}{n}\right)f\left(\frac{r}{n}\right)$

$=\lim_{n\to\infty}\frac{1}{n}\sum_{r=0}^{2n}\frac{1}{1+\left(\frac{r}{n}\right)}$, taking $\frac{1}{n}$ outside the sign of summation.

Now $\lim_{n\to\infty}\frac{1}{n}\sum_{r=0}^{2n}\frac{1}{1+\left(\frac{r}{n}\right)}$ is of the form $\lim_{n\to\infty}\frac{1}{n}\Sigma f\left(\frac{r}{n}\right)$, where $f\left(\frac{r}{n}\right)=\frac{1}{1+\left(\frac{r}{n}\right)}$. The limits of r in this summation are 0 to 2n.

When r = 0, $\frac{r}{n}=\frac{0}{n}=0$

and when r = 2n, $\frac{r}{n}=\frac{2n}{n}=2.$

As n → ∞, these values of $\frac{r}{n}$ tend to 0 to 2 respectively, giving us the limits of integration.

Now replacing $\frac{r}{n}$ by x, $\frac{1}{n}$ by dx, $\lim_{n\to\infty}\Sigma$ by the sign of integration $\int$, taking the limits of integration of x from 0 to 2, we get

$$\lim_{n\to\infty}\frac{1}{n}\sum_{r=0}^{2n}\frac{1}{1+\left(\frac{r}{n}\right)}=\int_0^2\frac{1}{1+x}dx$$

$$=\left[\log(1+x)\right]_0^2=\log 3-\log 1=\log 3-0=\log 3.$$

Example 73:

Evaluate the following:

$$\lim_{n\to\infty}\left[\frac{1}{n+1}+\frac{1}{n+2}+\ldots+\frac{1}{2n}\right].$$

Solution:

Here the general term (*i.e.*, the rth term) $=\dfrac{1}{n+r}$ and r varies from 1 to n. Thus we have to find $\lim_{n\to\infty}\sum_{r=1}^{n}\dfrac{1}{n+r}$.

We have $\lim_{n\to\infty}\sum_{r=1}^{n}\dfrac{1}{n+r}=\lim_{n\to\infty}\sum_{r=1}^{n}\dfrac{1}{\left\{1+\left(\dfrac{r}{n}\right)\right\}}$

$$=\lim_{n\to\infty}\frac{1}{n}\sum_{r=1}^{n}\frac{1}{1+\left(\dfrac{r}{n}\right)}.$$

The limits of r in this summation are 1 to n. Therefore the lower limit of integration $=\lim_{n\to\infty}\dfrac{1}{n}=0$,

and the upper limit of integration $=\lim_{n\to\infty}\dfrac{n}{n}=\lim_{n\to\infty}1=1$.

Hence the required limit

$$=\int_0^1\frac{1}{1+x}dx=\left[\log(1+x)\right]_0^1=\log 2.$$

Example 74:

Evaluate the following:

$$\lim_{n\to\infty}\left[\frac{1}{n+m}+\frac{1}{n+2m}+\ldots+\frac{1}{n+nm}\right].$$

Solution:

Here the rth term $=\dfrac{1}{n+rm}=\dfrac{1}{n}\left\{\dfrac{1}{1+\left(\dfrac{r}{n}\right)m}\right\}$ and r varies from 1 to n.

$\therefore$ the given limit $=\lim_{n\to\infty}\sum_{r=1}^{n}\dfrac{1}{n}\cdot\left\{\dfrac{1}{1+\left(\dfrac{1}{n}\right)m}\right\}$

$$= \lim_{n\to\infty} \frac{1}{n} \sum_{r=1}^{n} \frac{1}{1+\left(\frac{r}{n}\right)m}.$$

Also the lower limit of integration

$$= \lim_{n\to\infty} \left(\frac{1}{n}\right) = 0, \qquad [\because r = 1 \text{ for the first term}]$$

and the upper limit of integration $= \lim_{n\to\infty} \left(\frac{n}{n}\right) = 1$,

$[\because r = n$ for the last term$]$.

$\therefore$ the required limit $= \int_0^1 \frac{1}{1+mx} dx$

$$= \left[\frac{1}{m}\log(1+mx)\right]_0^1 = \left(\frac{1}{m}\right)\log(1+m).$$

Example 75:

Find the limit, when $n \to \infty$, of the series

$$\frac{n}{(n+1)^2} + \frac{n}{(n+2)^2} + \ldots + \frac{n}{(n+n)^2}.$$

Solution:

Here the rth term

$$= \frac{n}{(n+r)^2} = \frac{n}{n^2\left\{1+\left(\frac{r}{n}\right)\right\}^2}$$

$= \frac{1}{n} \cdot \frac{1}{\left\{1+\left(\frac{1}{n}\right)\right\}^2}$, and r varies from 1 to n.

$\therefore$ we have to find

$$\lim_{n\to\infty} \sum_{r=1}^{n} \frac{1}{n} \cdot \frac{1}{\left\{1+\left(\frac{r}{n}\right)\right\}^2}.$$

The lower limit of integration

$$= \lim_{n\to\infty} \left(\frac{1}{n}\right) = 0, \qquad [\because r = 1 \text{ for the 1st term}]$$

$\therefore$ the required limit $= \int_0^1 \frac{1}{(1+x)^2} dx = \left[-\frac{1}{(1+x)}\right]_0^1$

$$= -\frac{1}{2} - (-1) = -\frac{1}{2} + 1 = \frac{1}{2}.$$

Example 76:

Evaluate the following limits:

(i) $\lim_{n\to\infty}\left[\left\{\sqrt{(n+1)} + \sqrt{(n+2)} + \ldots + \sqrt{(2n)}\right\}/n\sqrt{n}\right].$

(ii) $\lim_{n\to\infty}\left[\frac{1}{(n+1)(n+2)} + \frac{1}{(n+2)(n+4)} + \frac{1}{(n+3)(n+6)} + \ldots + \frac{1}{6n^2}\right].$

Solution:

(i) The given limit

$$= \lim_{n\to\infty}\sum_{r=1}^{n}\frac{\sqrt{(n+r)}}{n\sqrt{n}} = \lim_{n\to\infty}\frac{1}{n}\sum_{r=1}^{n}\sqrt{\left(1+\frac{r}{n}\right)}$$

$$= \int_0^1\sqrt{(1+x)}\,dx = \left[\frac{(1+x)^{3/2}}{\frac{3}{2}}\right]_0^1$$

$$= \frac{2}{3}\left[(1+x)^{3/2}\right]_0^1 = \frac{2}{3}\left[2^{3/2} - 1\right] = \frac{2}{3}\left[2\sqrt{2} - 1\right]$$

(ii) The given limit

$$= \lim_{n\to\infty} n\sum_{r=1}^{n}\frac{1}{(n+r)(n+2r)}$$

$$= \lim_{n\to\infty}\frac{n}{n^2}\sum_{r=1}^{n}\frac{1}{\left(1+\frac{r}{n}\right)\left(1+\frac{2r}{n}\right)}$$

$$= \lim_{n\to\infty}\frac{1}{n}\sum_{r=1}^{n}\frac{1}{\left(1+\frac{r}{n}\right)\left(1+\frac{2r}{n}\right)} = \int_0^1\frac{1}{(1+x)(1+2x)}dx.$$

Let $\frac{1}{(1+x)(1+2x)} \equiv \frac{A}{1+x} + \frac{B}{1+2x}.$

Then $A = -1$, $B = 2$.

$\therefore$ the reqired limit

$$= \int_0^1 \left[\frac{-1}{1+x} + \frac{2}{1+2x} \right] dx$$

$$= [-\log(1+x) + \log(1+2x)]_0^1$$

$$= \left[\log\left(\frac{1+2x}{1+x} \right) \right]_0^1 = \log\frac{3}{2} - \log 1 = \log\frac{3}{2}.$$

Example 77:

Evaluate $\lim\limits_{n\to\infty} \left[\frac{n}{n^2} + \frac{n}{n^2+1^2} + \frac{n}{n^2+2^2} + \ldots + \frac{n}{n^2+(n+1)^2} \right].$

Solution:

Here the rth term $= \frac{n}{n^2+(r-1)^2}$. As the rth term contains (r – 1), we consider the (r + 1)th term.

The (r + 1)th term

$$= \frac{n}{n^2+r^2} = \frac{n}{n^2\left\{1+\left(\frac{r}{n}\right)^2\right\}}$$

$$= \frac{1}{n} \cdot \left\{ \frac{1}{1+\left(\frac{r}{n}\right)^2} \right\}, \text{ and r varies from 0 to n + 1.}$$

∴ the given limit

$$= \lim_{n\to\infty} \sum_{r=0}^{n+1} \frac{1}{n} \left[\frac{1}{1+\left(\frac{r}{n}\right)^2} \right].$$

Also the lower limit of integration

$$= \lim_{n\to\infty}\left(\frac{0}{n}\right) = \lim_{n\to\infty} 0 = 0, \qquad (\because r = 0 \text{ for the 1st term})$$

and the upper limit

$$= \lim_{n\to\infty}\left(\frac{n+1}{n}\right) = \lim_{n\to\infty}\left(1+\frac{1}{n}\right) = 1,$$

∵ r = (n + 1) for the last term.

$\therefore$ the required limit $= \int_0^1 \frac{1}{1+x^2} dx$

$$= \left[\tan^{-1} x\right]_0^1 = \tan^{-1} 1 = \frac{\pi}{4}.$$

Example 78:

Evaluate

$$\lim_{n\to\infty}\left[\frac{n}{n^2+1^2}+\frac{n}{n^2+2^2}+...+\frac{1}{2n}\right].$$

Solution:

Here the rth term $= \frac{n}{n^2+r^2}$, and r varies from 0 to n.

We get the required limit $= \frac{\pi}{4}$.

Example 79:

Evaluate

$$\lim_{n\to\infty}\left[\frac{n+1}{n^2+1^2}+\frac{n+2}{n^2+2^2}+...+\frac{1}{n}\right].$$

Solution:

Here the rth term

$$= \frac{n+r}{n^2+r^2} = \frac{1+\left(\frac{r}{n}\right)}{n\left\{1+\left(\frac{r}{n}\right)^2\right\}} = \frac{1}{n}\cdot\left\{\frac{1+\left(\frac{r}{n}\right)}{1+\left(\frac{r}{n}\right)^2}\right\},$$

and r varies from 1 to n.

$\therefore$ the given limit

$$= \lim_{n\to\infty}\sum_{r=1}^{n}\frac{1}{n}\cdot\left\{\frac{1+\left(\frac{r}{n}\right)}{1+\left(\frac{r}{n}\right)^2}\right\}$$

$$= \int_0^1 \frac{x+1}{x^2+1}dx = \int_0^1\left[\frac{x}{x^2+1}+\frac{1}{x^2+1}\right]dx$$

$$= \left[\frac{1}{2}\log\left(x^2+1\right)+\tan^{-1} x\right]_0^1 = \frac{1}{2}\log 2 + \frac{\pi}{4}.$$

Example 80:

Evaluate $\lim_{n\to\infty}\left[\frac{1}{n^3}\left(1+4+9+16+...+n^2\right)\right]$.

Solution:

Here the rth term $=\frac{1}{n^3}\left(r^2\right)=\frac{1}{n}\cdot\left(\frac{r}{n}\right)^2$, and r varies from 1 to n.

$\therefore$ the given limit $=\lim_{n\to\infty}\sum_{r=1}^{n}\frac{1}{n}\cdot\left(\frac{r}{n}\right)^2$

$$=\int_0^1 x^2dx=\left[\frac{x^3}{3}\right]_0^1=\frac{1}{3}.$$

Example 81:

Prove that

$$\lim_{n\to\infty}\left[\frac{1^2}{1^3+n^3}+\frac{2^2}{2^3+n^3}+...+\frac{n^2}{n^3+n^3}\right]=\frac{1}{2}\log 2.$$

Solution:

Here the rth term

$$=\frac{r^2}{r^3+n^3}=\frac{1}{n^3}\left\{\frac{r^2}{\left(\frac{r}{n}\right)^3+1}\right\}=\frac{1}{n}\cdot\left\{\frac{(r/n)^2}{(r/n)^3+1}\right\},$$

and r varies from 1 to n.

$\therefore$ the given limit

$$=\lim_{n\to\infty}\sum_{r=1}^{n}\frac{1}{n}\cdot\left\{\frac{\left(\frac{r}{n}\right)^2}{\left(\frac{r}{n}\right)^3+1}\right\}$$

$$=\int_0^1\frac{x^2dx}{x^3+1}=\left[\frac{1}{3}\log\left(x^3+1\right)\right]_0^1$$

$$=\frac{1}{3}\log 2-\frac{1}{3}\log 1=\frac{1}{3}\log 2.$$

Example 82:

Evaluate $\lim_{n\to\infty}\left[\frac{1}{n}+\frac{n^2}{(n+1)^3}+\frac{n^2}{(n+2)^3}+...+\frac{1}{8n}\right]$.

Solution:

Here the general term

$$= \frac{n^2}{(n+r)^3} = \frac{n^2}{n^3\left\{1+\left(\frac{r}{n}\right)\right\}^3} = \frac{1}{n}\cdot\frac{1}{\left\{1+\left(\frac{r}{n}\right)\right\}^3},$$

and r varies from 0 to n.

$$\therefore \text{ the given limit } = \lim_{n\to\infty}\sum_{r=0}^{n}\frac{1}{n}\cdot\frac{1}{\left\{1+\left(\frac{r}{n}\right)\right\}^3}$$

$$= \int_0^1 \frac{1}{(1+x)^3}dx = \left[-\frac{1}{2(1+x)^2}\right]_0^1 = -\frac{1}{8}+\frac{1}{2} = \frac{3}{8}.$$

Example 83:

If f(x) = f(a + x), prove that $\int_0^{na} f(x)\,dx = n\int_0^a f(x)\,dx.$

Solution:

We have $\int_0^{na} f(x)dx = \int_0^a f(x)dx +$

$$\int_a^{2a} f(x)dx + \int_{2a}^{3a} f(x)dx + ... + \int_{(n-1)a}^{na} f(x)dx. \qquad ...(1)$$

Now $\int_a^{2a} f(x)dx = \int_0^a f(a+t)dt,$

putting x = a + t and changing the limits

$$= \int_0^a f(a+x)dx, \qquad \text{[by prop. 1]}$$

$$= \int_0^a f(x)dx, \text{ since } f(a + x) = f(x), \text{ as given} \qquad ...(2)$$

Also $\int_{2a}^{3a} f(x)dx = \int_a^{2a} f(a+t)dt,$

putting x = a + t and changing the limits

$$= \int_a^{2a} f(a+x)dx = \int_a^{2a} f(x)dx$$

$$= \int_0^a f(x)dx. \qquad \text{from (2).}$$

Thus, we can show that each integral of the R.H.S. of (1) is equal to $\int_0^a f(x)dx$ and these being n in number, therefore we have $\int_0^{na} f(x)dx = n\int_0^a f(x)dx$.

Note: Similarly, we can prove that

$$\int_a^{na} f(x)dx = (n-1)\int_0^a f(x)dx.$$

Example 84:

Evaluate $\int_0^{\pi} \cos^{2n} x\, dx$.

Solution:

We have $\int_0^{\pi} \cos^{2n} x\,dx = 2\int_0^{\pi/2} \cos^{2n} x\,dx$,

$\left[\because \int_0^{2a} f(x)dx = 2\int_0^a f(x)dx \text{ if } f(2a-x) = f(x)\right.$

Here taking $f(x) = \cos^{2n} x$, we see that

$\left.f(\pi - x) = \cos^{2n}(\pi - x) = (-\cos x)^{2n} = \cos^{2n} x = f(x)\right]$

$$= 2\cdot\frac{(2n-1)(2n-3)\ldots 3.1}{2n(2n-2)(2n-4)\ldots 4.2}\cdot\frac{\pi}{2}, \text{ by Walli's formula}$$

$$= \frac{(2n-1)(2n-3)\ldots 3.1}{2^n\cdot n!}\cdot\pi.$$

Example 85:

Evaluate $\int_0^{\pi} \cos^6 x\, dx$.

Solution:

Let $f(x) = \cos^6 x$. Then

$f(\pi - x) = \cos^6(\pi - x) = (-\cos x)^6 = \cos^6 x = f(x)$.

Now $\int_0^{2a} f(x)dx = 2\int_0^a f(x)\,dx$, if $f(2a - x) = f(x)$.

$$\therefore \int_0^{\pi} \cos^6 x\,dx = 2\int_0^{\pi/2} \cos^6 x\,dx$$

$$= 2\frac{5.3.1}{6.4.2}\cdot\frac{\pi}{2}, \qquad \text{(by Walli's formula)}$$

$$= \frac{5\pi}{16}.$$

Example 86:

Evaluate $\int_0^{\pi} \sin^3 x\, dx$.

Solution:

Let $f(x) = \sin^3 x$.

The $f(\pi - x) = \sin^3 (\pi - x) = \sin^3 x = f(x)$.

Now $\int_0^{2a} f(x)\,dx = 2\int_0^a f(x)\,dx$, if $f(2a - x) = f(x)$.

$\therefore \int_0^{\pi} \sin^3 x\,dx = 2\int_0^{\pi/2} \sin^3 x\,dx$

$= 2\cdot\frac{2}{3.1}\cdot 1 = \frac{4}{3}$, (by Walli's formula.)

Example 86:

Evaluate $\int_0^{\pi} \theta \sin^3 \theta\, d\theta$.

Solution:

Let $I = \int_0^{\pi} \theta.\sin^3\theta\, d\theta$. ...(1)

Then $I = \int_0^{\pi} (\pi - \theta)\sin^3(\pi - \theta)\,d\theta$,

$\left[\because \int_0^a f(x)\,dx = \int_0^a f(a - x)\,dx, \text{ refer prop. 4}\right]$

$= \int_0^{\pi} (\pi - \theta)\sin^3\theta\, d\theta$. ...(2)

Adding (1) and (2), we get $2I = \int_0^{\pi}\left[\theta\sin^3\theta + (\pi - \theta)\sin^3\theta\right]d\theta$

$= \int_0^{\pi} (\theta + \pi - \theta)\sin^3\theta\, d\theta$

$= \int_0^{\pi} \pi \sin^3\theta\, d\theta = \pi\int_0^{\pi} \sin^3\theta\, d\theta$

$= 2\pi\int_0^{\pi/2} \sin^3\theta\, d\theta$, (by a property of definite integrals; prop. 6)

$= 2\pi\cdot\frac{2}{3.1}.1$, (by Walli's formula)

$= \frac{4\pi}{3}$.

$\therefore I = \frac{2}{3}\pi$.

Example 87:

Evaluate the following integrals:

(i) $\int_{-1}^{1} \frac{x^2 \sin^{-1} x}{\sqrt{(1-x^2)}} dx$

(ii) $\int_{-a}^{a} x\sqrt{(a^2 - x^2)}\, dx$,

(iii) $\int_{-1}^{1} \frac{x \sin^{-1} x}{\sqrt{(1-x^2)}} dx$

Solution:

(i) Let $I = \int_{-1}^{1} \frac{x^2 \sin^{-1}}{\sqrt{(1-x^2)}} dx$.

Put $\sin^{-1} x = t$,

so that $\left\{\frac{1}{\sqrt{(1-x^2)}}\right\} dx = dt$

and $x = \sin t$.

When $x = -1$,

$t = \sin^{-1}(-1) = -\frac{\pi}{2}$

and when $x=1$,

$t = \sin^{-1} 1 = \frac{\pi}{2}$.

$\therefore\ I = \int_{-\pi/2}^{\pi/2} t \sin^2 t\, dt$.

Now let $f(t) = t \sin^2 t$.

Then $f(-t) = (-t) = (-t) \sin^2(-t)$

$= -t(-\sin t)^2 = -t \sin^2 t = -f(t)$.

Therefore f(t) is an odd function of t.

$\therefore\ I = \int_{-\pi/2}^{\pi/2} t \sin^2 t\, dt = 0$. (Refer property 5)

(ii) Let $I = \int_{-a}^{a} x\sqrt{(a^2 - x^2)}\, dx$.

Let $f(x) = x\sqrt{(a^2 - x^2)}$. The $f(-x) = -x\sqrt{\{a^2 - (-x)^2\}}$

$$= -x\sqrt{(a^2 - x^2)} = -f(x).$$

Therefore f(x) is an odd function of x.

$$\therefore \ I = \int_{-a}^{a} x\sqrt{(a^2 - x^2)}\,dx = 0. \qquad \text{(Refer property 5)}$$

(iii) Let $I = \int_{-1}^{1} \frac{x\sin^{-1}x}{\sqrt{(1-x^2)}}dx.$

Proceeding as in part (i) of this question, we have

$$I = \int_{-\pi/2}^{\pi/2} t\sin t\,dt.$$

Let f(t) = t sin t. Then f (– t) = (– t) sin (– t)
= t sin t, so that f(t) is an even function of t.

$$\therefore \ I = 2\int_{0}^{\pi/2} t\sin t\,dt, \qquad \text{(Refer property 5)}$$

$$= 2\left[t(-\cos t)\right]_0^{\pi/2} - 2\int_0^{\pi/2} 1.(-\cos t)dt$$

$$= 2\times 0 + 2\int_0^{\pi/2}\cos t\,dt = 2[\sin t]_0^{\pi/2} = 2[1-0] = 2.$$

Example 88:

Prove without performing integration that

$$\int_{-a}^{2a} \frac{x\,dx}{x^2+p^2} = \int_{a}^{2a} \frac{x\,dx}{x^2+p^2}.$$

Solution:

We have

$$\int_{-a}^{2a} \frac{x\,dx}{x^2+p^2} = \int_{-a}^{a} \frac{x\,dx}{x^2+p^2} + \int_{a}^{2a} \frac{x\,dx}{x^2+p^2}. \qquad ...(1)$$

But if $f(x) = \frac{x}{x^2+p^2}$,

then $f(-x) = \frac{-x}{x^2+p^2} = -f(x)$.

Therefore f(x) is an odd function of x.

$$\therefore \ \int_{-a}^{a} \frac{x\,dx}{x^2+p^2} = 0.$$

So from (1), we get

$$\int_{-a}^{2a} \frac{x\,dx}{x^2+p^2} = \int_{0}^{2a} \frac{x\,dx}{x^2+p^2}.$$

Example 89:

Evaluate the following integrals:

(i) $\int_0^{\pi} \frac{dx}{a+b\cos x}$

(ii) $\int_0^{2\pi} \frac{dx}{a+b\cos x + c\sin x}$

Example 90:

Evaluate the following integrals:

(i) $I = \int_0^{\pi} \frac{dx}{a+b\cos x}.$ ...(1)

Then $I = \int_0^{\pi} \frac{dx}{a+b\cos(\pi - x)}$ (prop. 4) $= \int_0^{\pi} \frac{dx}{a-b\cos x}$ ...(2)

Adding (1) and (2), we get

$$2I = \int_0^{\pi} \frac{2a}{a^2-b^2\cos^2 x} dx$$

$$= 2a.2\int_0^{\pi/2} \frac{dx}{a^2-b^2\cos^2 x}. \qquad \text{(Refer property 6)}$$

$$\therefore\ I = 2a\int_0^{\pi/2} \frac{dx}{a^2-b^2\cos^2 x} = 2a\int_0^{\pi/2} \frac{\sec^2 x\,dx}{a^2\sec^2 x - b^2}$$

$$= 2a\int_0^{\pi/2} \frac{\sec^2 x\,dx}{a^2(1+\tan^2 x)-b^2} = 2a\int_0^{\pi/2} \frac{\sec^2 x\,dx}{a^2\tan^2 x + a^2 - b^2}.$$

Now let $a > b > 0$.

Put $a \tan x = t$,

so that $a \sec^2 x\ dx = dt$.

The new limits for t are 0 to ∞.

$$\therefore\ I = 2\int_0^{\infty} \frac{dt}{t^2 + \left\{\sqrt{(a^2-b^2)}\right\}^2}$$

$$= 2\frac{1}{\sqrt{(a^2-b^2)}}\left[\tan^{-1}\frac{t}{\sqrt{(a^2-b^2)}}\right]_0^\infty$$

$$= \frac{2}{\sqrt{(a^2-b^2)}}\left[\tan^{-1}\infty - \tan^{-1}0\right]$$

$$= \frac{2}{\sqrt{(a^2-b^2)}}\cdot\left(\frac{\pi}{2}-0\right) = \frac{\pi}{\sqrt{(a^2-b^2)}}.$$

(ii) Let $I = \int_0^{2\pi}\frac{dx}{a+b\cos x + c\sin x}$.

Let $b = r\cos\alpha$

and $c = r\sin\alpha$,

so that $r = \sqrt{(b^2+c^2)}$ and $\tan\alpha = \frac{c}{b}$.

Then $b\cos x + c\sin x = r(\cos\alpha\cos x + \sin\alpha\sin x)$

$= r\cos(x-\alpha)$.

$$\therefore\ I = \int_0^{2\pi}\frac{dx}{a+r\cos(x-\alpha)}.$$

Put $x - \alpha = t$,

so that $dx = dt$.

When $x = 0$, $t = -\alpha$ and

when $x = 2\pi$, $t = 2\pi - \alpha$.

$$\therefore\ I = \int_{-\alpha}^{2\pi-\alpha}\frac{dt}{a+r\cos t}$$

$$= \int_{-\alpha}^{0}\frac{dt}{a+r\cos t} + \int_0^{2\pi-\alpha}\frac{dt}{a+r\cos t}. \qquad ...(1)$$

(Refer property 3)

Now put $t = z - 2\pi$ in the first integral on the R.H.S. of (1). Then $dt = dz$ and the limits for z are $2\pi - \alpha$ to 2π.

$$\therefore\ I = \int_{2\pi-\alpha}^{2\pi}\frac{dz}{a+r\cos(z-2\pi)} + \int_0^{2\pi-\alpha}\frac{dt}{a+r\cos t}$$

$$= \int_0^{2\pi-\alpha}\frac{dt}{a+r\cos t} + \int_{2\pi-\alpha}^{2\pi}\frac{dz}{a+r\cos z}$$

$$= \int_0^{2\pi-\alpha} \frac{dt}{a + r\cos t} + \int_{2\pi-\alpha}^{2\pi} \frac{dt}{a + r\cos t},$$ because a definite integral does not change by changing the variable

$$= \int_0^{2\pi} \frac{dt}{a + r\cos t}, \qquad \text{(Refer property 3)}$$

$$= 2\int_0^{\pi} \frac{dt}{a + r\cos t}, \qquad \text{(Refer property 6)}$$

Now proceeding as in part (i) of this question, we get

$$\int_0^{\pi} \frac{dt}{a + r\cos t} = \frac{\pi}{\sqrt{(a^2 - r^2)}}, \text{ provided } a > r > 0.$$

$$\therefore \quad I = \frac{2\pi}{\sqrt{[a^2 - (b^2 + c^2)]}}, \text{ provided } a > \sqrt{(a^2 + c^2)} > 0$$

$$= \frac{2\pi}{\sqrt{(a^2 - b^2 - c^2)}}.$$